More Science Braintwisters and Paradoxes

By the same author

SCIENCE BRAINTWISTERS, PARADOXES, AND FALLACIES

More Science Braintwisters and Paradoxes

Christopher P. Jargocki

 VAN NOSTRAND REINHOLD COMPANY
NEW YORK   CINCINNATI   TORONTO   LONDON   MELBOURNE

Library of Congress Catalog Card Number: 82-4761
ISBN: 0-442-24524-6

Manufactured in the United States of America

Published by Van Nostrand Reinhold Company Inc.
135 West 50th Street, New York, N.Y. 10020

Van Nostrand Reinhold Publishing
1410 Birchmount Road
Scarborough, Ontario M1P 2E7, Canada

Van Nostrand Reinhold
480 Latrobe Street
Melbourne, Victoria 3000, Australia

Van Nostrand Reinhold Company Limited
Molly Millars Lane
Wokingham, Berkshire, England

15 14 13 12 11 10 9 8 7 6 5 4 3 2 1

Library of Congress Cataloging in Publication Data
Jargocki, Christopher P.
  More science braintwisters and paradoxes.

  Includes index.
  I. Science—Problems, exercises, etc. I. Title.
Q182.J36   793.73   82-4761
ISBN 0-442-24524-6   AACR2

DUNBAR BRANCH

To my late grandmother—Helena Jargocka

Preface

The book contains 194 puzzles based on scientific principles, each with a detailed solution. The problems deal with such topics as forces and motion, artillery, bicycles, sailboats, balloons, smoke rings, waves, siphons, parachutes, animals, trees, sounds and voices, musical instruments, heat, boiling and melting, magnets, electricity, rainbows, vision, mirrors, weather, rivers, mountains, rockets, satellites, weightlessness, the moon, planets, and many others.

As the title indicates, the puzzles are of three kinds. There are brain-twisters: although the lighted area of the full moon is only twice as large as that of the moon at first or last quarter, the full moon is about nine times brighter. Why? Paradoxes: when you stir a cup of tea, the leaves surprisingly collect in the center. Wouldn't you expect them to move outward like in a centrifuge? And there are fallacies: would it be possible to propel a sailboat on a windless day by mounting a large fan on the boat and directing the wind into the sails?

As these examples indicate, most of the puzzles contain an element of surprise. Indeed, the clash between commonsense conjecture and scientific reality is the central theme that runs through the book.

The puzzles range in difficulty from a few simple tongue-in-cheek questions to subtle problems requiring a lot of thought. Most puzzles are nonmathematical and require only a qualitative application of broad physical principles. All are designed to reward the reader with a great deal of physical intuition and insight into the world around him. If one can understand these knotty problems, the rest of classical physics should become much easier.

The book can be read with profit by high school and college students, science teachers, and all puzzle enthusiasts.

I am deeply indebted to Dr. Franklin Potter of the University of California at Irvine with whom I had many stimulating discussions, and who read and reviewed the original manuscript. My special thanks go to Prof. Myron Bander, my Ph.D. thesis adviser, and Prof. Meinhard Mayer, both of the University of California at Irvine, who were kind enough to review the first draft of the book. I am also grateful to the UCI Physics faculty and graduate students, particularly Prof. Riley Newman, Dr. Peter Amendt, Dr. Kerry Hoskins, Dr. Robert Kares, and Dr. Toshi Tachi who participated in the seminars I based on the present volume.

I am especially grateful to my mother, Stefania Vcala, for encouragement and great help in locating certain invaluable references. Last but not least, I want to offer my greatest thanks to Terri Corneth who did a superb job in typing the original manuscript and created an atmosphere of warmth and emotional support without which this book could not have been written.

C. P. Jargocki

Contents

More Science Braintwisters and Paradoxes

QUESTIONS

1. Forces and Motion

1. Dick and Jane race each other in a 100-yard dash. Dick wins by 10 yards. They decide to race again, but this time to give Jane an equal chance, Dick begins 10 yards behind the start line. Assuming that both run with the same constant speed as before, who is the winner this time?

2. Try this experiment at home. Fasten two straws together at one end with paper clips. Place the straws together over a third straw or pencil, balance a marble at the lower end, and gradually separate the straws (see diagram). Surprisingly, the marble will roll toward the upper end! How can the marble appear to defy gravity?

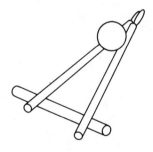

3. What will happen if two identical cannons are aimed directly at each other and the bullets fired simultaneously and at the same speeds (see diagram)? One cannon is higher than the other but the two are perfectly aligned.

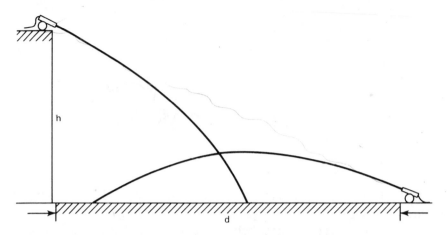

4. If you shake a bucket partially filled with apples of various sizes for a couple of minutes, the biggest apples will always end up on top. Why?

5. Generally speaking, bodies with low centers of gravity are more stable than those with high. For example, a stub of a pencil can be stood on its end very easily, but it is much harder to stand a long stick on its end. Paradoxically, however, a long stick with its high center of gravity is much easier to balance on the tip of a finger than a short pencil. Why?

6. Newton's law of gravitation is sometimes expressed by the equation

$$F = Gm_1m_2/d^2$$

where F represents the force between two masses m_1 and m_2, d is the distance between centers of mass, and G is a constant. Is this a correct formulation of Newton's law of gravitation?

7. There is a popular toy consisting of five steel balls, all of the same size and mass, hanging side by side in a row (see diagram). Pull out the end ball and drop it against the row, and one pops off the other end. If two are pulled aside and dropped together, two pop out from the other end. The balls can obviously count! How do they accomplish this trick?

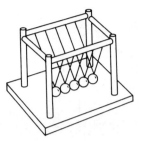

8. Imagine a ball of mass m moving with a speed v, and colliding head on with a massive wall. If the collision is elastic, the ball simply recoils with the same speed v. But if this is true, then the kinetic energy of the ball, $mv^2/2$, is conserved, but its momentum mv is not, because its velocity is now directed in the opposite direction.

The thoughtful reader may say that the laws of the conservation of momentum and energy should be applied to the total system consisting of the ball and the wall (or wall + earth). The momentum change of the ball, $mv - m(-v) = 2mv$, will be equal to the momentum change of the wall + earth, $MV - M \cdot 0 = MV$. But then the energy will not be conserved for the total energy before the collision is $mv^2/2$ and the total energy after the collision is $mv^2/2 + MV^2/2$. What's the way out of this paradox?

9. Which weighs more, a cubic meter of large-sized coal or a cubic meter of small-sized coal? Assume that all the pieces of coal in each cubic meter are loosely packed identical-size spheres, each sphere touching six others.

10. A ball lies on the floor touching a wall which makes an obtuse angle with the floor (see diagram). We can resolve the weight of the ball into two components: perpendicular to the wall and parallel to the floor. By Newton's third law of motion, there is a reaction of the wall on the ball counterbalancing the component of its weight perpendicular to it. But then the component of the weight parallel to the floor will remain unbalanced, and the ball will have to have a horizontal acceleration. However, the ball is totally unmoved by our argument and just lies waiting for us to find an error in our reasoning. Where is that error?

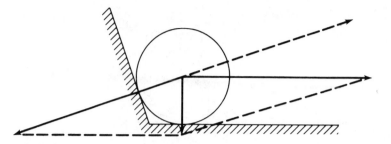

11. A bucket is left out in the rain. Will the rate at which the bucket is filled with water be changed if a wind starts to blow?

12. A spring balance is hung from the ceiling by means of a long rope. A second rope is attached to the spring balance, pulled tight so the balance reads 100 pounds, and then anchored to the floor (see diagram). If a 60-pound weight is hung on the hook of the balance, what will the balance read?

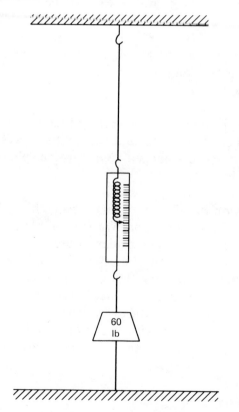

13. Consider a contraption shown in the diagram, which consists of a block of wood to which is attached a long bent rod bearing two heavy metal spheres. Why doesn't it fall off the table?

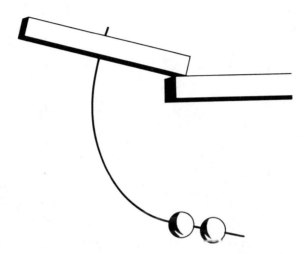

14. Set up four inclined planes to form a rhombus (see diagram). Now release two identical balls simultaneously from A so one rolls along ABC, and the other along ADC. Which ball do you think will reach the bottom first?

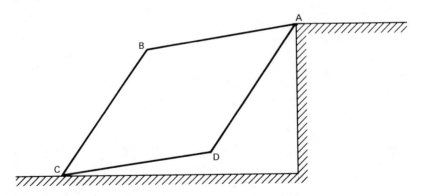

15. The center of mass CM of an isosceles triangle is located at a point one third up the altitude of the triangle above the base (see part [a] of the diagram). Now consider a right circular cone of the same cross section (see part [b] of the diagram). Is its center of mass also located at a point one third up the altitude of the cone?

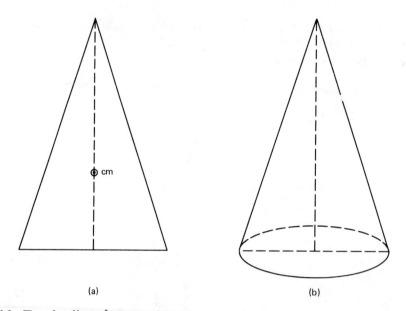

(a)   (b)

16. Two bodies of masses M and m are raised to the same height above the floor, and released simultaneously. Assuming that the air resistance is constant and identical for the two bodies, will they reach the floor simultaneously?

2. Man-Made World

1. In very long-range artillery the guns are often set at an elevation angle ranging from 50 to 70°, instead of the 45° angle which, according to elementary mechanics, is supposed to guarantee the maximum range. Why?

2. You have two cylinders of identical size and mass. They are made of two materials of different densities. One of the cylinders is hollow, How can you tell which one?

3. Balance a stick horizontally on the index fingers of both hands. Now slide your fingers together. No matter what the initial position of your fingers was, the stick will still balance. How can you explain that?

4. Why is it easier to drive a stake with a heavy hammer (even swung gently) than with a light one, although the latter can be swung with great speed thus giving it enormous energy?

5. Two identical pulleys with centers at the same level are connected by a belt (see diagram). The pulley on the left is the drive pulley. Is the maximum power that can be transmitted by the belt greater when the pulleys rotate clockwise or counterclockwise?

6. Here's an experiment you can try at home. Take a spool, preferably a big one like the kind that wire cable comes wound on. Wrap a wide ribbon around the axle of the spool so that it comes off the bottom. Now try pulling on the ribbon. Amazing things happen! By increasing the angle between the ribbon and the vertical the spool can be made to roll towards you, while by decreasing this angle it is made to roll away from you. A value of the angle can be found where the spool simply revolves and remains roughly in the same spot. How can you explain this strange behavior of the obedient spool?

7. A bicycle is being lightly held in a vertical position with the cranks vertical. A horizontal backward pull is applied to the lower pedal by a man standing on the ground. In what direction will (a) the bicycle start to move and (b) the crank rotate?

8. A student is holding a bicycle wheel with a lead-filled tire in front of his chest. He holds one end of the horizontal axle in each of his outstretched hands. The vertical wheel is set spinning between his arms as shown in the diagram. Suppose that the student wishes to make the plane of the spinning wheel rotate slightly to the left about its vertical axis? (That is, have the axle remain horizontal while its left end moves closer to his ribs and its right end moves farther away.) Will pushing forward with his right hand and backward with his left do the trick?

9. Can the coefficient of static friction exceed unity? In other words, can the frictional force between a body and a surface be greater than the weight of the body?

10. Steel girders used in construction often have the form of I beams. Their cross section (see diagram) shows that most of the material is collected in large flanges at the top and bottom, whereas the web joining the flanges is rather thin. Why is this peculiar shape so universal?

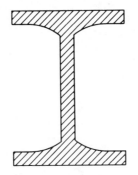

11. Suppose you fire a bullet from a rifle tilted at 45° to the ground. According to elementary textbooks this angle gives a maximum range if we neglect air resistance. By how much do you think the maximum range will be reduced if we do take air resistance into account?

12. Imagine two bridges which are exact replicas of each other except that every dimension in one is five times as large as in the other, i.e., the first bridge is five times longer, its structural members are five times thicker, etc. Which bridge is stronger?

13. Imagine riding a bicycle along a straight path when, suddenly, due to unevenness in the road or a wind gust you find yourself tilting to one side. A beginner will instinctively try to steer to the other side, and soon he will have bruises to show for it. In contrast, an experienced cyclist steers into the direction of the fall. Why?

14. When a bicycle is suddenly tilting to one side, one gets out of this predicament by steering into the direction of the fall. In contrast, when a cyclist is about to turn a corner, just before reaching it he will first throw the front wheel over in the opposite direction. Why?

15. If the walls of an exceptionally tall or wide building are exactly vertical, can they be made precisely parallel?

3. Gases

1. The diagram shows a dime placed 1–2 centimeters from the edge of a table, and a cup tilted so that the lip is about 2 centimeters off the top of the table, and about 2–3 centimeters behind the dime. Can you get the dime off the table and into the cup without touching the coin?

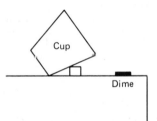

2. Imagine two sailboats built to exactly the same proportions, except that one is twice as large as the other, i.e., its masts are twice as thick, and its sails twice as long and twice as wide, although they are made of the same kind of canvas. Which sailboat will be more likely to have its sails torn under the force of wind?

3. Why are industrial smokestacks generally very tall?

4. Would it be possible to propel a sailboat on a windless day by mounting a large fan on the boat and directing the wind into the sails?

5. Galileo didn't believe in the existence of the atmospheric pressure. To support his view he would give the following reasoning. Imagine a container with a liquid. Within the liquid consider a certain volume. The liquid inside this volume is acted upon by two oppositely directed forces: weight and buoyant force. According to Archimedes' Principle the two forces are equal. This is why the liquid is in equilibrium; it neither sinks nor floats upward. We might say, for example, that water immersed in water is weightless. How then can something weightless exert pressure on the underlying layers!

Similarly air in air, being weightless, cannot exert pressure on the lower layers, and ultimately on the surface of the earth. What is the flaw in his reasoning?

6. Two soap bubbles, unequal in size, are blown on a T-shaped tube. The blowing inlet is then closed, leaving the bubbles connected (see diagram). What will happen?

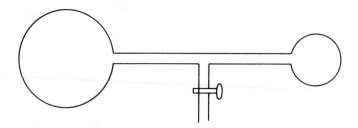

7. Parachutes are designed to slow down the fall. If so, why do they often have holes in the center?

8. Can an iceboat travel faster than the speed of the wind which propels it?

9. There are four communities $A, B, C,$ and D situated at the corners of a square of length a. It is decided to connect these towns by a highway system in such a way that any two towns are connected by a highway and the total length of the highway system is minimized. What should be the configuration of the highways?

10. Why do racing sailboats generally have curved sails rather than flat sails?

11. Which raindrops fall faster—large or small?

12. We all know that a smoke ring in still air travels slowly in a direction perpendicular to the plane of the ring (see diagram). In such a ring the smoke particles rotate around the hollow circular axis of the doughnut in the directions indicated with arrows. What makes smoke rings travel through the air? Will the smoke ring in the diagram travel to the left or to the right?

13. Have you ever seen the game of chase played by two smoke rings moving in the same direction? The trailing ring accelerates and shrinks, whereas the leading ring slows down and expands (see diagram). The smaller ring catches up with the larger one and passes through it. The roles are then reversed; the trailing ring becomes the leading one and vice versa, and the process is repeated. It is a fascinating show to watch, but how do we explain it?

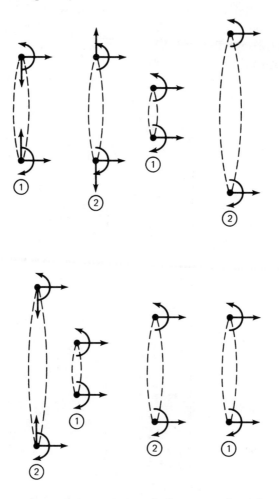

14. As we know the extra pressure at the bottom of a column of gas of unit area is just equal to the weight of the molecules in the column. Some people might say that the extra pressure is due to the increased speed of the molecules as they fall from top to bottom. Namely, if a molecule starts from rest and falls freely from height h, its velocity is given by $v = \sqrt{2gh}$. If the molecule simply bounced from the bottom surface, the change of momentum would be $\triangle(mv) = 2mv$. The impulse delivered to the bottom is $F\triangle t = \triangle(mv)$. If we use as the time interval between bounces the time of fall of the molecules through the distance h and back up again, the time can be found from $h = (1/2)gt^2$. Therefore

$$t_{\text{total}} = \triangle t = 2\sqrt{2h/g}$$

$$F = \triangle(mv)/\triangle t = \frac{2mv}{2\sqrt{\frac{2h}{g}}} = \frac{2m\sqrt{2gh}}{2\sqrt{\frac{2h}{g}}} = mg$$

For N molecules, the extra force on the bottom of the column is equal to Nmg, which is the weight of the molecules in the column.

Do you agree with this argument?

4. Liquids

1. Can you carry water in a sieve? To prove that this is possible, take a wire sieve with holes no smaller than 1 millimeter in diameter, and dip it into melted paraffin. The sieve will still have holes in it, but now it will be covered with a thin barely discernible film. Carefully pour some water into it. The water will not drip through! How do you explain this strange behavior?

2. When you stir a cup of tea, the leaves surprisingly collect in the center. Wouldn't you expect them to move outward like in a centrifuge?

3. One pan of a balance carries a container with water, and the other a stand with a weight suspended from it (see diagram). The pans are in balance. Then the stand is turned around so the suspended weight is completely submerged in the water. Obviously the balance is disturbed since the pan with the stand becomes lighter. What additional weight must be put on this pan to restore equilibrium?

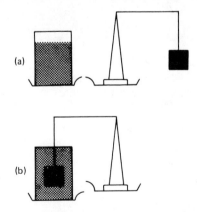

4. Is there any difference between waves in the sea and waves in a stream?

5. What is the mass of a cubic meter of Oriental soup?

6. A container holds the equivalent of 30 glasses of water. If you turn on the tap at the bottom, it takes 10 seconds to fill a glass with water. How long will it take before the container runs dry if the tap is left turned on?

7. A man on the shore is pulling in a boat by taking in a rope attached to its bow at a constant speed v. The speed of the boat, v_b, will then apparently be given by the horizontal component of v, namely $v_b = v\cos\alpha$. Thus the greater the angle α (i.e., the closer the boat is to the shore), the slower the speed at which the boat is being pulled in. This, however, contradicts everyday experience as can be seen by pulling in a pencil with a piece of thread—the pencil will move faster. What's then wrong with our reasoning?

8. Fill a glass with water almost to the top. Get a small cork stopper or any small object that will float high in the water. Now the problem is to place the cork in the water and have it stay floating in the middle of the glass. The cork may be placed anywhere on the water surface, and there are no other restrictions.

9. Let a drop of detergent or a flake of soap fall between two floating matches placed about a half-inch apart in a bowl of water. The matches will fly apart as though pulled by something. Why?

10. A bucket of water containing a cork held down on the bottom is dropped from the top of a building, the cork being released by some means at the moment the bucket is dropped. Where is the cork right before the bucket reaches the ground?

11. Suppose you are in a boat floating in a swimming pool. There is a stone lying on the bottom of the boat. You pick it up, and throw it overboard. Will the water level rise, fall, or remain unchanged?

12. A finger placed in a foaming glass of beer will cause the "head" to settle down. Why?

13. Why is it generally easier for women to float on their back than for men?

14. Is it easier to make bubbles using a liquid of high surface tension or low surface tension?

15. (Due to John B. Hart) The diagram shows a coiled garden hose which behaves in a very strange way: If you pour water into the upper end of the hose, none will come out the other end. What's even more surprising, very little water will enter the hose. Why?

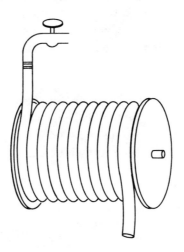

16. Three identical glasses *A, B,* and *C* are arranged as shown in the diagram. *A* and *B* are filled with water, which can best be done by submerging them, and putting the mouths of the glasses together. *B* is supported on *C* by a few hollow stirrers. Some additional hollow stirrers are available on the table. Now, the problem is to transfer the water from glass *A* to glass *C* without touching or moving the glasses or the hollow stirrers supporting glass *B*. The additional hollow stirrers may be moved but may not touch the glasses or the stirrers supporting *B*.

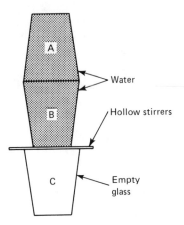

17. When a drop of salt water falls into a glass of clear water, you can see the vortex ring created by the splash. Since salt water is heavier than plain water, the ring is pulled downward by a net gravitational force greater than buoyancy. If this force is roughly constant, we should see the vortex ring move down with increasing speed. Instead, the ring is observed to be *slowing down* and increasing in size in its downward course. In fact, this is true in general: the larger the external force on a vortex ring, the slower the motion of the ring through a fluid. Can you explain this paradoxical behavior?

18. The siphon has been used since ancient times for getting liquids up over the edge of a container and into another one at a lower level. Despite this everyday familiarity many people consider the operation of the siphon somewhat mysterious. For example, some think that the liquid is pushed over the siphon by air pressure. However, siphons can operate in a vacuum. So how do siphons really work?

19. Why is it that a siphon doesn't start itself, but rather must be started by suction?

20. Some books say that the maximum elevation (uptake) over which a siphon will work is given by the height of a liquid that can be supported

by the atmospheric pressure less a correction for friction in the tube, i.e., about 9.0–9.6 meters for water or 0.71–0.74 meters for mercury. However, we showed in the preceding puzzles that atmospheric pressure is not responsible for the working of the siphon. Then why is it so hard, though possible, for the uptake to exceed the barometric height?

5. Living World

1. In parachute jumping the effective landing speed is often equivalent to a jump from a second story window (~ 10 feet). If you are not careful, are you more likely to break your neck or snap your ankles?

2. Does the pig's habit of wallowing in mud serve any useful purpose?

3. The African elephant has very large ears. From a physicist's point of view, could they play a useful role in the animal's hot habitat?

4. Surface tension is a force which is hardly noticeable to large animals, and yet is deadly to insects. Why?

5. Man consumes about one fiftieth of his own weight in food daily, but a mouse will eat half its own weight in a day. Thus, per pound of body weight, mice eat 25 times as much food as people, and yet don't seem to grow any bigger. So where does it all go?

6. An adult cat's tail tends to bend over when held erect. On the other hand, a kitten's tail can easily stand up spiky and straight. Why?

7. When measuring blood pressure the inflatable cuff is always wrapped around the upper arm. Why?

8. If you drop a cat upside down, it will always mysteriously land on its feet. How can the animal achieve a net rotation in space without having anything to push against?

9. How do long balancing poles help tightrope walkers maintain their balance?

10. When a pirouetting skater pulls in her arms, she turns much faster. The angular momentum remains constant since there is no torque about the skater's vertical axis. Thus $L = I_1\omega_1 = I_2\omega_2$, where I and ω are the moment of inertia and the angular velocity, respectively. Now what about the energy? The kinetic energies before and after are $E_1 = 1/2\ I_1\omega_1^2 = 1/2\ L\omega_1$ and $E_2 = 1/2\ I_2\omega_2^2 = 1/2\ L\omega_2$. Since $\omega_2 > \omega_1$ the kinetic energy after is greater than before! Where did the extra energy come from?

11. Can a bicyclist go faster than 100 miles per hour on level ground?

12. The maximum running speed on level ground seems to be independent of the size of an animal. For example, a rabbit can run as fast as a horse. However, in running uphill small animals easily out-pace larger ones: A dog can easily run up a hill while a horse slows its pace. These findings can be easily demonstrated with simple dimensional arguments. Can you see how?

13. Is there any physical advantage to the vee formation often assumed by migrating birds?

14. A tree must transport nutrients between its central trunk and outermost leaves along a reasonably direct path. Why then can't a tree sustain each of its leaves with a separate branch? In other words, why is the branching pattern shown in part (a) of the diagram so much more common in nature than the explosive pattern in part (b)?

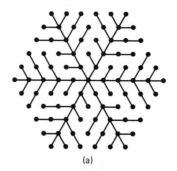

(a)

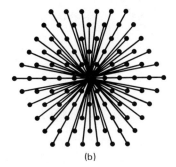

(b)

15. If you look carefully at a tree, you will observe that small branches generally stick out at almost right angles from big ones and big branches stick out at smaller angles from each other. Any idea why?

16. Leonardo da Vinci made the following observation regarding trees and their branches: "All the branches of trees at every stage of their height, united together, are equal to the thickness of their trunk." This seemingly plausible rule turned out not to be confirmed by detailed studies. Can you see that the rule violates the principle of least effort?

6. Sound

1. In most cases it is easier for a male speaker to fill a room with his voice than for a female speaker. Why?

2. Sound generally travels much faster in liquids and solids as compared to gases. For example, in steel the velocity of sound is about 5050 m/sec; in sea water it is around 1500 m/sec, and in air around 340 m/sec. On the other hand, lead has a sound velocity of only 1200 m/sec, and rubber, even more surprisingly, of only 62 m/sec! Why these exceptions?

3. Why do foghorns emit very low-pitched sounds?

4. Take a long piece of wood, and put your ear to one end. Stretch out your arm, and scratch the most distant place on the wood that you can reach. Your scratching will sound quite loud, yet if you take your ear away and go on scratching as before, there is hardly a sound to hear. Why?

5. When you "crack" your knuckles, what actually produces the cracking sound?

6. Why do people's voices sound high-pitched when they inhale helium?

7. Why is it that in a piano the wires are struck with hammers covered with soft felt?

8. The diagram shows two concert hall designs differing in the shape of the ceiling above the orchestra. The numbers show the time differences between the times of arrival of the direct and reflected sound. Which concert hall will have better acoustics?

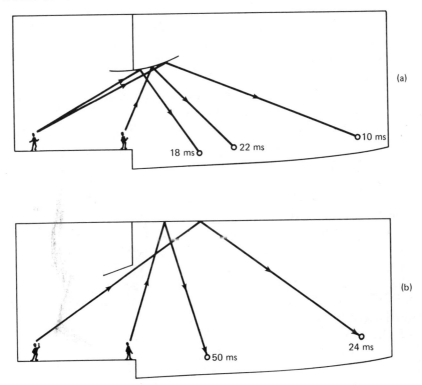

(a)

18 ms   22 ms   10 ms

(b)

50 ms   24 ms

9. The speed of sound in air is about 760 mph. Suppose now that a police car is sounding its siren, and going toward you at 80 mph. Is the sound of the siren coming toward you at 760 mph + 80 mph = 840 mph, as measured relative to the ground?

10. Which is the preferable location for a factory housing noisy machinery: on a hilltop or in a valley?

11. Sometimes when we listen to someone talking over a public address system, we find that his voice is almost drowned out by squeals and howls from the loudspeakers. What is happening?

12. We have all heard about a mouse that roared, but is there any scientific basis for it? On the other hand, could an elephant emit a high-frequency squeak?

13. If you are a hi-fi lover, you know that for maximum listening pleasure you have to readjust the bass-treble controls of your equipment every time you change the volume at which you listen to your records. Why can't the tone controls be set once and for all?

14. One needs a long resonating pipe or a large cavity to produce bass tones. How is it then possible for telephones and small transistor radios to reproduce bass tones? They all have tiny speakers, yet we do hear bass from them.

15. Echoes, as we know, are reflections of the sound waves from distant objects. It would seem that in mountainous areas where the terrain is more varied, there would be plenty of opportunity for hearing echoes. Paradoxically, the opposite is the case: Echoes occur much more seldom and are harder to catch. Why?

16. On a well-tuned piano, despite its pleasing sound, each note is actually slightly out of tune with the others. In fact, if the piano were tuned perfectly, it would grate on your ear and in places seem drastically out of tune. What's the explanation?

7. Heat

1. Some people claim that they can briefly plunge their hands into molten lead and escape injury. Is there any scientific basis for these boasts?

2. In his book *Prairie* James Fenimore Cooper describes a method of fighting a prairie fire by setting a counterfire. Despite the fact that the prairie wind was blowing toward the travelers and bringing the fire closer, the counterfire would move in the opposite direction away from the travelers. How do you explain this seeming paradox?

3. A medical thermometer has a constriction in the tube that prevents the mercury from falling as soon as you remove the thermometer from your mouth. However, the mercury did go once through the constriction when the temperature was being taken. Why won't it go back?

4. Place a large jar over a burning candle which is sitting in a pan of water. You will observe that the flame soon goes out and the water rises dramatically, perhaps even flooding the candle. Why?

5. To prevent floods aircraft sometimes drop black soot on snowed-in mountain sides a few weeks before the beginning of spring. Why should this work?

6. Consider communicating vessels connected by a narrow tube with a stopcock. Assume that at first all the liquid is in the left vessel and its height is h. Then we open the stopcock and the liquid flows from the left into the right vessel. After some sloshing the flow finally stops when there is an equal level of liquid, $h/2$, in each vessel. Let us

calculate the initial and final potential energy of the liquid. Initially the potential energy was $Wh/2$, i.e., weight times the height of the center of gravity. At the end it changed to $(W/2)(h/4) + (W/2)(h/4) = Wh/4$—half of the initial potential energy disappeared! What happened to it?

7. For complete comfort, cooling the air by air conditioning should always be accompanied by its dehumidification. Why?

8. A door and the metal doorknob on it are obviously at the same room temperature. Yet the doorknob feels colder to touch than the door. Does it mean that our senses are incapable of judging the temperature of objects correctly?

9. Would it be possible to cool a room by leaving the refrigerator door open?

10. If you turn on the heater in your room, and after, say, an hour you turn it off, will the total energy of the air in the room be raised by the heating?

11. You have a metal ring with the central hole 10 inches in diameter. The ring is made red hot, expanding in every direction. Does the central hole get smaller?

12. Before pouring tea into a glass, it may help to put in a teaspoon to prevent the glass from cracking. Why should this trick work?

13. Is it possible to show that ice doesn't melt in boiling water? Very easily. Take a test tube and fill it with water. Now take a piece of ice and press it down on the bottom of the tube with a small weight. If you now heat the test tube so that the flame licks only the upper part of the tube, the water will soon be boiling, but oddly enough the ice at the bottom just won't melt! How does one explain this paradoxical behavior?

14. Two identical spheres have the same temperature. One of them is suspended by a string, and the other lies on a table. Both spheres receive identical amounts of heat. We assume that the heat is trans-

ferred so fast that none is lost to surrounding objects. Will the spheres have the same or different temperatures upon the addition of heat?

15. Will turning out incandescent lights save energy in winter? How about in summer?

16. Can you boil water in boiling water? For example, take a small bottle, fill it with water, and suspend it in boiling water so it doesn't touch the bottom. If you keep the bottle in boiling water long enough, will the water inside it come to boil?

17. Would it be possible for ice made from ordinary water to be so hot that it would burn your fingers?

18. In 1714 Fahrenheit chose as the zero point of his scale the lowest temperature then obtainable in the laboratory: -17.7°C. This point was reached using a mixture of water, granulated ice, and ammonium chloride. Paradoxically, even though the temperature of such a mixture drops by almost 18°C, the amount of heat energy in it remains unchanged (assuming that none has flown in from the surroundings). How is this possible?

19. If a drop of water falls on a moderately hot stove, it spreads out and quickly evaporates. But if the stove is very hot, the water gathers up into a little ball which may dance around for as long as a minute or two before evaporating. What explains this paradoxical behavior known as the spheroidal state?

20. Two wooden pails, without lids, containing hot and cold water are set out in freezing weather. Which pail will begin to freeze first?

8. Electricity and Magnetism

1. Place a burning candle between the oppositely charged poles of a high-voltage generator of static electricity. The flame will be mysteriously attracted toward the negatively charged pole and repelled from the positive pole. Any explanation?

2. Why are "keepers" used on permanent magnets?

3. Occasionally one still sees gasoline trucks dragging chains beneath them. What is the reason for this practice?

4. Imagine a transformer in which the secondary coil has four times the number of turns of wire as the primary, yet is made from wire having one fourth the resistance per unit of length. In this case the two coils will have equal resistances. Moreover, the secondary would seem to have four times as much voltage and four times as much current. This would clearly violate the law of conservation of energy. Is there a way out of this paradox?

5. You have an unmagnetized iron bar but there is no magnet around. What is a simple way of magnetizing and demagnetizing the bar?

6. In a hydrogen atom the negatively charged electron is circling around the positively charged proton. Unlike charges are strongly attracted to each other, so why doesn't the electron fall toward the proton? Note that you can't answer: "For the same reason that the earth doesn't fall toward the sun.'' The notion of an orbit has no meaning in the atomic world so the answer must involve different concepts.

7. Why would you be quite safe using a compass to find the direction of true north in Vermont, but not in British Columbia?

8. Banks of capacitors are sometimes installed on low-voltage power lines served by a local transformer station. What is their purpose?

9. As we know like stationary charges repel and unlike stationary charges attract. However, like currents attract and unlike currents repel. This seems paradoxical because an electric current is also composed of charges, although in motion. Why is there such a striking difference in behavior between stationary charges and moving charges?

10. AM antennas on cars and portable transistor radios are usually vertical. Why?

11. You are aboard an airplane flying in the midst of a thunderstorm. Any moment the plane might be hit by lightning. Are you in any danger?

12. If you are inside a hollow conductor, you are completely shielded from any outside electric charges or fields. Let us now reverse the situation: Put a metal shield around a charge. If you are on the outside, will you detect any electric forces due to the charge inside?

13. There is a trick that physicists like to play on their unsuspecting layman friends. They place a bar magnet between vertical wooden posts, then let go, and presto! The magnet hangs in the air, seemingly defying gravity. Of course, what the laymen don't know is that there is another magnet hidden in the wooden base. When the like poles are on the same sides, the magnets repel. However, when the floating magnet is turned end for end, repulsion is replaced by attraction, and the top magnet crashes (see diagram). Now, the question is: Is it possible to make magnets so that the upper magnet will float *regardless* of which way it is oriented with respect to the lower one?

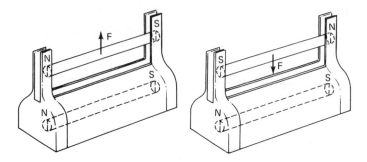

14. As we know, two parallel wires carrying currents in the same direction attract each other. This is what we mean when we say "like currents attract." Now, suppose that we have two parallel electron beams like those used in cathode-ray tubes. If the electrons move in the same direction, will the beams attract or repel each other?

15. We know that a conducting housing provides protection from outside electrostatic fields (see Puzzles 11 and 12 this chapter). An obvious question comes to mind: Is it likewise possible to prevent a magnetic field from entering an enclosure?

16. A charged particle of charge q and velocity v moving in a magnetic field of intensity B has a force F on it given by $F = qv \times B$. By definition of the cross product the force is always perpendicular to the velocity. Thus a magnetic field can never do any work on charged particles. However, in a lab one can easily demonstrate that a current-carrying wire placed in a magnetic field will gain kinetic energy and start moving. How can this be if the field cannot do any work on the charges moving in the wire?

17. The Kelvin water dropper is an amazing contraption that can generate voltages up to 15,000 volts (see diagram). Cans A and D as well as cans B and C are connected electrically with wire. Water drips through the two bottomless tin cans A and B, and is collected in cans C and D. In a few seconds one connected pair of cans becomes positively charged, while the other pair becomes negatively charged. The voltages are so high that a small neon bulb brought close to one of the cans will flash. What makes this ingenious device work?

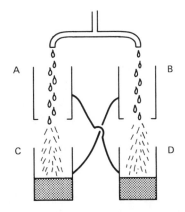

18. The energy of a capacitor is given by $E = Q^2/2C$, assuming the charges $+Q$ and $-Q$ reside on the capacitor plates, and its capacitance is C. Suppose that we increase the distance d between the plates of an isolated parallel-plate capacitor. Since C is inversely proportional to d, the capacitance will decrease. The energy E, on the other hand, will increase as Q remains unchanged. This makes physical sense: The plates attract, being oppositely charged. To increase their separation, we must do work which goes into increasing the energy of the capacitor. Now do a different experiment: Connect the capacitor between the terminals of a battery, and again increase the separation between the plates. The energy can be reexpressed as $(1/2)CV^2$, where V is the voltage difference between the plates. In this experiment V remains constant as we increase the separation, C decreases and therefore the energy also decreases. This would seem to indicate that here the capacitor does work in helping the external agent move the plates farther apart. It is as if in this case the oppositely charged plates repelled each other. How can this be?

19. If you ever rode in a car equipped with both an AM and FM radio, you may have noticed that while the AM radio cuts out when you are passing under a bridge, the FM radio in the same situation would continue to play. Why is there such a big difference in the reception of AM and FM signals?

9. Light and Vision

1. A rainbow can only be seen in the morning or late afternoon. Why?

2. How can you explain the following paradox: an iron bar heated to 800°C glows brightly, but a piece of quartz heated to the same temperature will hardly glow at all!

3. Try the following experiment. Turn on a flashlight and shine the beam into a glass tank full of water. The light beam will be seen to bend sharply *downward* at the point where it enters the water. Now put a straight stick into the water at an angle. The part in the water will seem to bend *upward*! Why the contradiction?

4. Even though animals differ greatly in size, their eyes seem to have similar dimensions. As a result the eyes of small or young animals look proportionately larger than our own. Why?

5. Why is your image in a plane mirror inverted left-to-right but not top-to-bottom?

6. Few people realize that glasses are not the only way to correct poor vision. Make a pinhole in an opaque material and place it over the eye so that you can look through the pinhole. For example, a pinhole in a pop bottle cap serves this purpose well. When placed over eyes suffering from nearsightedness or farsightedness a remarkable improvement in vision occurs! What is the optical basis for this pinhole monocle?

7. On clear days the sky appears blue. However, on hazy, foggy, or cloudy days the sky appears white. Why the difference?

8. Have you ever wondered why the human eye detects certain wavelengths of light and not others? In otherwords why are the wavelengths from 0.00040 millimeter for extreme violet to 0.00077 millimeter for extreme red visible?

9. A window pane viewed at night from inside a lighted room acts as a mirror, and we can see our image in it. Why does this only happen at night?

10. Is it possible to make a mirror that does not reverse left and right?

11. A radiometer consists of four vanes free to rotate about a pivot inside a glass bulb containing air at very low pressure (see diagram). One side of each vane is blackened and readily absorbs light; the other side is silvered and reflects most light. The vanes rotate when they are illuminated.

Sometimes one hears the following explanation of this phenomenon: Light consists of particles called photons. When a photon strikes the blackened side, it is absorbed. (That's why the surface looks black.) In the process the surface receives the photon's momentum p. On the other hand, when a photon hits the silvered side, it rebounds from it with a reversed momentum $-p$. The momentum is conserved, so the white surface must receive twice the photon's momentum, $2p$, since then the initial momentum p will be equal to the total final momentum $-p + 2p$. Thus, the black sides get the energy and heat up faster, but the white ones experience twice as much radiation pressure. As a result, the vanes will rotate in the direction toward which the black sides are directed.

Do you agree with this explanation?

12. Suppose you put a radiometer into a cold and dark place, for example, a refrigerator. Will it spin, and if so in which direction?

13. How is it possible for Polaroid sunglasses to eliminate the glare without blocking out the rest of the world?

10. Spaceship Earth

1. Do you agree with the following homespun weather predictions? If you do, what is the scientific basis for them?

(a) Your joints are more likely to ache before a rainstorm;

(b) Frogs croak more before a storm;

(c) If leaves show their undersides, rain is due;

(d) A ring around the moon means rain if the weather has been clear;

(e) Birds and bats fly lower before a storm;

(f) You can tell temperature by listening to a cricket;

(g) Ropes tighten up before a storm;

(h) Fish come to the surface before a storm;

(i) "Singing" telephone wires signal a change in the weather.

2. In Miami during a clear June night the temperature may fall to only 68°F. from a high of 86°F. However, at El Paso, Texas the temperature could fall to as low as 50°F by morning. Why such a big difference?

3. An observer on the beach always sees the larger waves come in directly toward him even though some distance out from shore they are seen to be approaching at an angle. What makes the waves straighten out?

4. Residents of skyscrapers have often observed that on a clear winter night the ground temperature may get close to 0°F while on top of the skyscraper it may be as warm as 32°F. How can we explain this?

5. Every stationary iron object in the United States is magnetized with a north pole at the bottom and a south pole at the top! This includes bathtubs, filing cabinets, refrigerators, and even umbrellas with an iron shaft left standing for a while. Any ideas why?

6. The temperature always rises a little when it rains or snows. Why?

7. When an airliner is flying at an altitude of 30,000 feet, the temperature of the air outside may be as low as −30°F. However, instead of heaters, air conditioners must be used in aircraft flying that high. Why?

8. Lost polar explorers are reported to have a strong tendency to circle steadily toward the right near the North Pole, and to the left near the South Pole. Can you think of a possible explanation?

9. Winds on the earth blow directly from higher-pressure areas to lower-pressure ones. True or false?

10. The diagram shows the successive positions of the sun during a period of a few hours, as observed in Alaska. Can you tell approximately which compass direction was the observer facing? Roughly, at what time of day or night was the lowest elevation of the sun observed?

11. There is no such thing as a straight river. In fact it was found that the distance any river is straight does not exceed 10 times its width at that point. At first we might suppose that a river twists and bends in direct response to peaks and dips in the landscape. Not at all! On a typical smooth and gentle slope water does *not* flow straight downhill; it winds and turns desperately trying to avoid the straight path to the bottom. Why?

12. A common unit of time in astronomy is the solar day. Paradoxically, in the northern hemisphere solar days are longer in winter than they are in summer. Can you see how this comes about?

13. The theory of continental drift is today widely accepted among geologists. In particular, they say that the present distribution of land masses makes our epoch extremely prone to glaciation. Why?

14. For purely astronomical reasons the southern hemisphere of the earth should suffer colder winters and hotter summers than its northern counterpart. In fact, the lowest temperature ever recorded, $-126.9°F$, occurred in the Antarctic. However, by and large the peculiar conditions existing in the southern hemisphere compensate this trend very effectively. What mysterious astronomical reasons and peculiar conditions are we referring to?

15. When cold and warm air lie alongside each other, as they do in a weather front, even if no pressure difference exists at ground level, the warm air and the cold air will act as high and low pressure zones, respectively. The pressure difference between them then gives rise to the so-called thermal winds. On the other hand, we know that cold air is heavier than warm, so it seems that it is the cold air that should be associated with a high-pressure zone. How do we resolve that contradiction?

16. You might expect the gravitational attraction due to a nearby mountain range to cause a plumb bob to hang at an angle slightly different from vertical. This example actually appears in many physics textbooks. However, the observed deflection is, surprisingly, much smaller than what is predicted by theoretical calculations. In fact, the deflection is practically zero, apparently implying that a mountain range exerts no extra pull on a plumb bob. Can you see a way out of the paradoxical situation?

17. Two identical trains travel at exactly the same speeds, in opposite directions, east to west, and west to east. Assuming they are traveling along the same parallel, which of the trains is heavier?

18. Upon arriving in Mexico, visitors from Canada are often surprised at how fast it gets dark after sunset as compared to the relatively long period of twilight in northern latitudes. Why the difference?

11. Space Exploration

1. Why were the first American satellites launched from Cape Canaveral, Florida?

2. Can a satellite be launched so that it will hover over, say, New York, 24 hours a day?

3. Suppose you are in a windowless room aboard a wheel-shaped space station. The station is spinning about its hub to maintain normal simulated gravity. What simple test can you make to convince yourself you are aboard a space station, and not on earth?

4. If a rocket is launched vertically upward with a speed of 11.2 km/sec, it will be able to escape from the earth. Now suppose that the rocket is launched almost horizontally with the same speed. Neglecting air resistance, will the rocket still be able to escape from the earth?

5. Imagine a base on the moon with streets and sidewalks just like on the earth. Would walking be easier or harder on the moon?

6. When a satellite separates from the launching rocket that was used to put it in orbit around the earth, the rocket is usually seen to overtake the satellite gradually even though its motor has been shut off. Any ideas why?

7. Why is it that for a flight to Venus space vehicles are aimed backward along the orbit of the earth, whereas for a flight to Mars they are fired in the direction that the earth is moving?

8. When Skylab was in orbit around the earth, the astronauts aboard Skylab lived in a state of weightlessness, and yet their weight was carefully recorded every day. How is it possible to weigh something that has no weight?

9. Imagine a rocket moving parallel to the ground high above the earth. Is it possible for the exhaust gases to move in the same direction as the rocket with respect to the ground, and still accelerate the rocket forward?

10. In elementary mechanics we learn that when you throw a stone, it follows a parabolic arc (neglecting air resistance). However, the rocket experts will tell you that a projectile will not follow a parabola unless it was fired at a speed equal to or greater than the so-called *escape* or *parabolic speed*. The latter is equal to 11.2 km/sec ($\approx$ 7 mi/sec). In this case the projectile will never return to the earth. They will also say that an object fired at less than the escape speed will follow an elliptical path with the earth at one of the foci. Who is right?

11. Weightlessness can be achieved for a period lasting up to a minute in an airplane executing only one of the following maneuvers: (1) Inside loop (with the center above the plane); (2) Outside circular loop (with the center below the plane); (3) Outside parabolic loop. Which one?

12. During lift-offs and reentries astronauts always assume a prone (i.e., parallel to the ground) position. Why is this preferable to a sitting-up position?

13. The first manned landing on the moon made by the crew of Apollo XI on July 16, 1969, as well as the subsequent five landings all took place on the side of the moon facing the earth. Why?

12. The Universe

1. Is the Earth in any danger of falling into the sun?

2. Although the lighted area of the full moon is only twice as large as that of the moon at first or last quarter, the full moon is about nine times brighter! Why?

3. Can you see the "earthrise" or "earthset" on the moon?

4. The inner edge of Saturn's rings revolves 2 1/2 miles per second faster than the outer edge. Based on this information, would it be possible for Saturn's rings to be thin solid disks?

5. Venus and the earth are about the same size. However, viewed from Venus, the earth at its best would appear about six times brighter than Venus ever appears to the Earth. This is despite the fact that the earth is farther away from the sun! How can you explain this paradox?

6. Why are Mercury and Venus generally invisible at night?

7. On any clear night a meteor can be seen in the sky about every 10 minutes. However, their number increases toward morning. Why?

8. The mean density of the Earth is 5.5 g/cm³, i.e., 5.5 times that of water. On the other hand, the four massive planets of our solar system have much lower densities: Neptune—2 g/cm³; Uranus—1.5 g/cm³; Jupiter—1.34 g/cm³; and Saturn only 0.69 g/cm³. What is the reason for this difference?

9. Are there any natural objects in the solar system that rise in the west and set in the east?

10. The planets of our solar system show a very interesting relationship between mass and period of rotation. In general, the greater the mass, the faster the speed of rotation. Thus, Jupiter, with the largest mass of any of the planets, also has the fastest speed of rotation: 9 hours, 50 minutes. Saturn with a smaller mass rotates in 10 hours, 15 seconds. Uranus and Neptune, with masses still smaller, rotate in 11 or 12 hours. Finally, Mars, which is far smaller than any of the outer planets, rotates in 24 hours, 37 minutes. However, the earth is 10 times as massive as Mars, yet it rotates in about the same time. Why does the Earth rotate so slowly?

11. The flags of a number of countries (e.g., Comoro Islands, Mauritania, Pakistan) have stars between the horns of the crescent moon. Can this ever be seen in the sky?

12. The moon, like the sun, often appears in different hues of yellow and red. But can the moon ever appear blue, as in the expression "once in a blue moon"?

13. The tallest mountain on Earth is not Mount Everest, but the Hawaiian volcano Mauna Kea. It rises 31,000 feet from the ocean floor, surpassing Mount Everest by nearly 2,000 feet, but only the top 13,823 feet show above the surface.

Surprisingly, however, the tallest mountain on Mars, volcanic cone Olympus Mons, is at least 80,000 feet high and its base is 350 miles in diameter. Mars is only about half the earth's size, and yet its mountains are much taller than ours. Any explanation?

14. Do the people in the Southern Hemisphere see the moon upside down?

15. What is the *weight* of the moon?

16. We all know that the moon appears larger when it is close to the horizon. Does the same effect happen with stars? In other words, do the constellations "expand" as they approach the horizon?

ANSWERS

1. Forces and Motion

1. Dick is the winner again. In the first race he ran 100 yards in the time it took Jane to run 90. Therefore, in the second race, after Jane has gone 90 yards, Dick will have gone 100, so he will be alongside her. Both will have 10 more yards to go. Since Dick is the faster runner, he will finish before Jane.

2. Actually, the center of mass of the marble falls slightly as the marble rolls between the widening straws.

3. The surprising answer is that no matter what the distance between the cannons is and at what angle they are aimed, the bullets will always collide in flight (neglecting air resistance).

To understand why, let us for the moment turn off the force of gravity. The bullets will then follow the straight line shown in the diagram, and collide at its midpoint. Now, bring back the force of gravity. Instead of following the straight line, the bullets fall from it by *equal* distances, so will still collide in midair.

4. A system assumes its most stable position when its potential energy reaches a minimum. The center of gravity of the apples will be in its lowest position when the apples in the lower portion of the bucket become as densely packed as possible. This will be the case if all the nooks and crannies in the lower portion are filled with small apples. As a result, the larger apples will tend to end up on top.

5. The rule that bodies with low centers of gravity are more stable than those with high applies only to situations involving static equilibrium. When balancing a stick on the tip of a finger, the finger is continuously moved so as to keep it underneath the center of gravity of the stick. This is because to assure balance the point of support must always be beneath the center of gravity. However, when the stick is long, its moment of inertia, i.e., its resistance to turning, relative to the point of support, is large. As a result the force of gravity which is trying to turn the falling stick even farther away from the vertical can only do it very slowly, thus giving the person enough time to shift the finger back underneath the center of gravity of the stick.

6. Consider a carpenter's square as shown in the diagram. The center of mass of this square is at the point C. Now place a small spherical body at C. The distance between the centers of mass of these two bodies, the sphere and the carpenter's square is zero! According to the statement of Newton's law of gravitation given in the question the attractive force between the two bodies would become infinite, which is clearly not the case. Indeed one may conclude that if the small sphere is placed at A the attractive force may tend to increase the distance between the centers of mass. Thus in terms of the statement of the law given in the question the force becomes repulsive rather than attractive!

Newton was well aware of the problem of determining the distance to be used in the formula for the gravitational force. In his formulation the inverse-square law applies to mass particles rather than extended bodies, and the distance d is the distance between two mass particles. It is only in the special case of spheres whose densities are dependent only on the distance from their centers that d refers to the distance between their centers of mass, i.e., their geometrical centers.

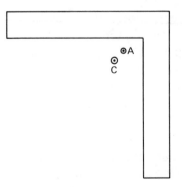

7. The toy illustrates the principles of conservation of momentum and energy. Suppose two right-hand balls are released from rest at a height h and strike the other balls with a velocity v. The total momentum just before the collision is therefore $2mv$. After the collision the three right-hand balls are stationary, and two left-hand balls fly away with velocity v and total momentum $2mv$, exactly equal to the total momentum just before the collision. The energy is also conserved because the two left-hand balls rise to exactly the same height h at which the two right-hand balls were released. Thus the final potential energy of the two left-hand balls, $mgh + mgh = 2mgh$, is equal to the initial potential energy of the two right-hand balls, $mgh + mgh$.

There is a common misconception that the principles of conservation of momentum and energy are enough to predict the behavior of the balls. This is not true. The two principles give us only two equations for the unknown final velocities. Even if only three balls were used, two equations in three unknowns would result! To get a solution we must bring some other principle into play. For example, we can assume that the balls are separated by a small distance and each collision involves only two neighboring balls. Then in each case we only have to find two final velocities. When two elastic balls of the same mass collide head on, they simply exchange their velocities. In our case, if the end ball on the right strikes its stationary neighbor with a velocity v, the end ball will come to rest, while its neighbor to the left will start moving with the velocity v. In this fashion the momentum is transmitted along the row. Using this principle we can work out more difficult problems, for example one in which the end ball of mass $2m$ strikes three balls of mass m each. Upon collision, the velocity of the mass $2m$ will be $v/3$ and that of the mass m will be $4v/3$, based on the equations for conservation of momentum and kinetic energy. Similar calculations can be made for each collision along the row, and the problem will be completely solved.

8. The ball must rebound from the wall with a speed slightly less than the incident speed, even though the collision is elastic. This effect is well known for gas molecules striking a movable piston, and thus doing work during the expansion of the gas. The momentum of the wall + earth, $P = MV$, is a finite quantity equal to $2mv$. The kinetic energy of the wall + earth can be expressed as

$$K = 1/2\,MV^2 = P^2/2M.$$

If P is finite, and M, which is the combined mass of the wall and the earth, becomes very large, the kinetic energy tends to zero. This shows that a massive object may have appreciable momentum while at the same time having practically zero kinetic energy. Both quantities are then conserved.

9. The weight of the coal is the same in both cases. Some people think that the weight of the small-sized coal would be greater because of its tighter packing. What they overlook is that there are far more voids in the case of the smaller sized coal, although each void is smaller.

Let us prove our claim. If n is the number of spheres along a side 1 meter long, then the total number of spheres in the cubic meter is n^3, and the radius of each sphere is $1/2n$ meter. The volume of a sphere is $(4/3)\pi r^3$. So the volume of one of these spheres is $(4/3)\pi\,(1/2n)^3 = (1/6)\pi/n^3$ meter^3, and as there are n^3 spheres, the total volume of the spheres in 1 meter^3 is $\pi/6 \approx 0.5236$ meter^3. Thus, slightly more than half is coal, and slightly less than half is air space. Since this fraction is independent of the radius of each sphere or their number, it must be identical for both large and small spheres.

10. Newton's third law of motion should not be applied separately to each *component* of a real force such as the gravitational force with which the earth attracts the ball. In fact, Newton's third law should not be applied to the *resultants* of real forces, either.

11. The rate of filling, i.e., the amount of rain that falls into the bucket per second, will not change! Although the cross-sectional area of the rain falling into the bucket decreases ($S_1 = S\cos\alpha$, see diagram), the velocity of the raindrops not only changes direction but also increases in magnitude ($v_1 = v/\cos\alpha$, see diagram). In other words, the rate at which the bucket fills up depends only on the vertical component of the velocity of the raindrops, which is not altered by the wind.

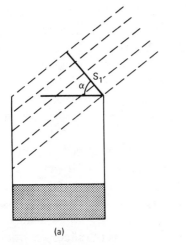

(a)

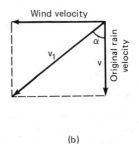

(b)

12. The balance will read 100 pounds! When the 60-pound weight was hung on the hook of the balance, the tension in the lower rope immediately decreased to 100 lb − 60 lb = 40 lb. However, the sum of the downward forces exerted by the 60-pound weight and the 40-pound tension in the rope still added up to 100 pounds. The 60-pound weight took away some of the burden carried by the rope, but the total burden remained the same. Thus if any weight up to 100 pounds is hung on the hook, the reading will remain 100 pounds. If a 100-pound weight is hung on it, the tension in the rope will become zero, the weight having taken over the role of the rope. If more than 100 pounds are hung on the hook, the rope will become completely slack, and the reading will be equal to that of the weight suspended from the hook.

13. The center of gravity of the contraption is located in the region of the two spheres. Draw an imaginary line connecting it to the point of support. You will find that the center of gravity lies directly beneath the point of support, and any deflection of the rod bearing the two spheres to the left or right results in the raising of the center of gravity. Thus the equilibrium is stable. We conclude that it doesn't matter if the center of gravity lies directly beneath or above the point of support for an object to be in equilibrium. What is important, however, is that they lie on the same vertical line.

14. The ball rolling along *ADC* will reach the bottom first. True, the balls do cover identical distances, and the accelerations along *AB* and *DC* as well as those along *AD* and *BC* are the same because of the same inclinations of the planes. However, the ball traveling along *DC* will have a high average velocity acquired during its rapid descent along *AD*. On the other hand, the ball traveling along the corresponding side *AB* will have a very small average velocity since its acceleration is small.

15. The center of mass of a right circular cone is located at a point one *fourth* up the altitude of the cone. The reason for this lowering of the center of mass will become clear if we imagine the cone to be composed of thin triangular slices parallel to the largest triangular slice passing through the vertex. The center of mass of each such triangular slice lies one third up the altitude of the slice. However, as the slices get smaller toward the outside of the cone, so do the altitudes, and the centers of mass get closer and closer to the base of the cone. As a result, the center of mass of the whole cone is brought down to a point one fourth up its axis.

16. One might be tempted to say that since the air resistance is the same for the two bodies, it can be disregarded. Consequently, the two bodies will reach the floor at the same time. However, as we can easily see this is wrong. Let us take, for example, the body of mass M. It is subject to two forces: the weight Mg and the air resistance F. The resultant force is $Mg - F$. From this the acceleration is $a = (Mg - F)/M = g - F/M$. Hence the body with the larger mass has a higher acceleration, and will reach the floor first.

2. Man-made World

1. When a shell is fired at a steep angle with a great initial velocity, it reaches the altitude of 25–30 miles where the air is very rarefied and offers little resistance. The shell may then fly 80–100 miles through the stratosphere before veering steeply back to the earth. Were the shell to be fired at a 45° angle, its trajectory would run through the dense atmospheric layers, limiting its range to only a few miles.

2. By letting them roll without slipping down an inclined plane. At the bottom of the plane the total kinetic energies of the cylinders must be equal since they fell through the same height. The total kinetic energy T consists of the translational part and the rotational part

$$T = mv^2/2 + I\omega^2/2$$

where I is the moment of inertia and ω is the angular velocity. However, $v = a\omega$, where a is the radius of the cylinder. Thus

$$T = \omega^2(ma^2 + I)/2$$

The equality of T's for the two cylinders can now be written as

$$\omega_s^2(ma^2 + I_s)/2 = \omega_h^2(ma^2 + I_h)/2$$

where s and h stand for "solid" and "hollow," respectively. But $I_h > I_s$ since the mass of the hollow cylinder is farther away from the axis of symmetry. Therefore, $\omega_h < \omega_s$, i.e., the hollow cylinder will roll slower than the solid one.

3. The stick will be seen to slide first on one finger, and then on the other, switching back and forth until both fingers come together underneath the stick's center of gravity. This strange behavior is explained by noting that the finger which is farther away from the stick's center of gravity carries a lighter load. Hence the force of friction there is less, and so this finger will start moving first. As the moving finger is brought closer to the center of gravity, more and more of the stick's weight will be on that finger until the kinetic friction between it and the stick becomes greater than the static friction between the other finger and the stick. Then the first finger will stop, and the second finger will begin to slide. The roles alternate several times until both fingers reach the stick's center of gravity.

4. How much kinetic energy a hammer has is not important. What matters is how much of that energy can be transferred to the stake. The collision between the hammer and the stake may be compared to the collision between two equal-sized balls. When a rolling metal ball hits a stationary wooden ball head on, the latter will start rolling at a low speed while the metal ball will continue rolling in the same direction. When a rolling wooden ball hits a stationary metal ball, the latter will hardly move whereas the wooden ball will rebound with almost the same speed. In both cases very little energy has been transferred from the moving ball to the stationary one. On the other hand, if a rolling metal ball strikes an identical stationary ball head on, the first ball will stop dead in its tracks and the second ball will take off with a speed equal to that of the first ball. In this case all of the kinetic energy of the moving ball was transferred to the stationary ball. The rule that emerges from these experiments is that for maximum energy transfer the masses of the colliding objects should be as close as possible. Thus a heavy hammer being closer in mass to the stake can transfer more of its energy to it than a light hammer.

5. The sagging of the working part of the belt will be less when the pulleys rotate clockwise. As a result the belt will wrap itself around a greater portion of the pulleys' perimeters, increasing the coupling between them and the power transmitted.

6. The forces acting on the spool are shown in the diagram. In the static case the horizontal and vertical force components as well as the torques about point A at the center of the spool are all zero. This gives us the following three equations

$$Fsin\theta = f = \mu N,$$
$$N + Fcos\theta = W,$$
$$Fr = fR,$$

where f is the frictional force between the spool and the table, N is the normal force exerted by the table on the spool, μ is the coefficient of sliding friction, W is the weight of the spool, r is the smaller radius about which the ribbon is wrapped, and R is the larger radius in contact with the table. Combining the first equation with the third, we get

$$sin\theta = r/R$$

which shows that the value of θ for which the spool remains in the same spot is independent of the coefficient of friction. Now, from the second equation $f = \mu N = \mu(W - Fcos\theta)$. Thus if θ is made larger than its equilibrium value, $cos\theta$ decreases making both the term in the parentheses and the force of friction f larger. As a result the torque fR prevails over Fr, and the spool rotates clockwise, i.e., toward the experimenter. If θ is made smaller, the torque Fr prevails over fR, and the spool rotates counterclockwise, i.e., away from the experimenter.

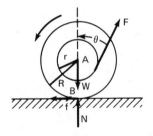

7. The bicycle will move backwards and the crank will rotate clockwise! Looking at the diagram, we note the following relationships: The resultant torque acting on the crank is zero, i.e.,

$$Tr_2 - Fr_1 = 0 \qquad\qquad (2-1)$$

where T is the tension in the chain, and F is the applied force.

The resultant torque acting on the rear wheel is also zero, i.e.,

$$Tr_3 - Sr_4 = 0 \qquad\qquad (2-2)$$

where S is the forward force exerted on the rear wheel by the ground. From Equations $(2-1)$ and $(2-1)$ we get

$$S = Tr_3/r_4 = Fr_1r_3/r_2r_4$$

But for any normal bicycle $r_1 < r_4$ and $r_3 < r_2$, so $S < F$. As a result there is a net backward force $F - S$ acting on the bicycle frame. The bicycle will therefore move backwards, with the wheels and the crank rotating clockwise.

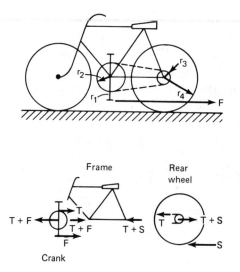

8. The answer, paradoxically, is that the student must push *upward* with his right hand and *downward* with his left.

To see why let us consider the state of motion of four mass elements in the tire. One at the top will have its velocity vector horizontal and straight away from the student's nose, and will require a small velocity change to the left to give the required effect. One at the bottom will have its velocity vector horizontal and straight toward his stomach, requiring a small change to the right. The ones at the front and back will have vertical velocity vectors, down and up, which require no changes at all for the desired shift of the wheel's orientation.

The argument can be extended readily, by taking horizontal and vertical velocity components, to show that all mass elements in the upper half of the wheel need a velocity change to the left while all those in the lower half need a velocity change to the right.

Since the only way to change the velocity of a mass element in a given direction is to apply a force in that same direction, it follows that the student must apply forces acting to the left on the upper half of the wheel and to the right on the lower half. He can do this through the axle, bearings, hub, and spokes, by pushing upward with his right hand and downward with his left.

9. It is a common misconception that the coefficient of static friction cannot exceed unity. Actually, as shown in the table below, it can be much greater than unity:

MATERIALS	μ	ANGLE OF REPOSE, θ
Dry tire on dry road	1.0	45°
Aluminum on aluminum	1.5	56°
Styrofoam on Styrofoam	2.1	65°
Plastic foam sponge on sandpaper	29	88°

The coefficient of static friction is often determined by placing a block of, say, aluminum on, say, an aluminum board, and then tilting the board until a limiting angle is reached beyond which the block will begin to slide down the board. This limiting angle is called the *angle of repose,* and for aluminum on aluminum it is equal to 56°. At this angle the component of the weight of the block along the board is just equal to the force of friction that tries to prevent the block from sliding down the board. It can be easily shown that the coefficient of friction is equal to the tangent of the angle of repose, i.e., $\mu = tan\ \theta$.

10. Take a wooden beam and support its ends on two chairs. If you now hang a few weights along the length of the beam, the result will be as shown in the diagram. The top layer of the beam is shortened, and has to resist compression, while the bottom layer is lengthened, and has to resist tension. Somewhere between the top and the bottom is a layer *mn*, which remains of the same length, and is therefore of no use except to connect top and bottom together. This is called the neutral layer. Steel is more expensive than wood, and much heavier, so when girders are made from it most of the material should be placed where it does most good. That's why there should be as much material as possible in the top and bottom layer, and as little as possible in and near the neutral layer.

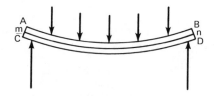

11. As the diagram shows, the air drag makes the range almost 10 times shorter! In the absence of air resistance a bullet flying with an initial velocity of 2000 ft/sec would describe a vast arc 6 miles high and fly almost 24 miles. But in reality the bullet only flies about 2.5 miles, describing a tiny arc compared to its first trajectory.

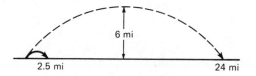

12. The smaller bridge is five times as strong as the larger bridge! The reason is that a bridge is as strong as its steel girders. The strength of a steel girder is proportional to its cross section which in turn varies as the square of the girder's linear dimension. But the weight of the girder is proportional to the cube of its linear dimension. Hence the load-carrying ability of a bridge varies inversely with its linear dimension. Thus if we take a bridge and increase its every dimension five times, the resulting large bridge will be so weak that it might even collapse under its own weight.

13. By steering into the direction of the fall the cyclist follows a curved path of such a radius as to generate enough centrifugal force to bring himself and the bicycle upright again. Once in the vertical position the cyclist turns the handlebars over to get closer to the original direction. Before this maneuver the cyclist and the rest of the bicycle had usually swung into line behind the front wheel because of a castering effect (see puzzle 14 this chapter). Also, the cyclist often might oversteer. In either case, to get out of the initial curve the cyclist is forced to go into another curve on the other side of the original direction. Thus he progresses by a series of arcs which at high speed become almost imperceptible. After all, at high speed the cyclist can afford the luxury of a large radius of curvature r, i.e., an almost straight path, since he gets a sufficient centrifugal force from the large v^2 term in the numerator of $F_{centrif} = mv^2/r$.

14. When we have to turn a corner, the curvature of the path could generate enough centrifugal force to knock us over to the outside of the turn. To counteract that, we have to lean into the turn so that the resultant of gravity and the centrifugal force will lie in the (tilted) plane of the bicycle. To get the necessary tilt, we all subconsciously throw the front wheel over to the outside of the turn. The resulting centrifugal force immediately tips us over toward the corner. As soon as we have the necessary tilt, we bring the handlebars over into the turn, and then easily round the corner. To get out of the turn, we turn even more sharply into the bend. This throws the bicycle toward the vertical. The moment we are upright again, we simply straighten out the front wheel, and once more follow a straight path.

There is one important element that we have overlooked so far. Today in all bicycles the front fork is not straight, as it still was in the modern-looking "Rover" safety bicycle introduced in 1885 by Starley, but rather is bent forward (see diagram [a]). As a result, the point at which the front tire meets the ground is behind the steering axis. Suppose now that the bicycle leans over (see diagram [b]). The force of gravity F acting on the front wheel is applied to the center of the wheel, and remains vertical at all times. When the bicycle leans over, F will make an angle with the plane of the wheel, and we can then decompose it into its components: $F_{\parallel}$, lying in the plane of the wheel, and $F_{\perp}$, perpendicular to it. Now, if the center of the wheel, O, were on the steering axis (i.e., if the fork were straight), then $F_\perp$ couldn't twist the wheel about the steering axis since its arm of action would be zero, and so the twisting torque would vanish. However, in the modern design there is a distance OA separating the point of application of the force, and the steering axis about which the front wheel turns. $F_\perp$ thus develops a torque which twists the front wheel in the direction of the lean. This effect is precisely what makes "no-hands" riding possible. Moreover, it is independent of the speed. Even if we tilt a stationary bicycle, its front wheel will obediently twist in the direction of the tilt.

We see that the bicycle by virtue of its design "helps" the rider in going into a curve, should he inadvertently tilt too much to one side. This, however, puts us at once in a dilemma. If it is so easy for the front wheel to twist in the direction of the tilt, then what's stopping the wheel from twisting all the way to the side? Some people hypothesized that the bicycle tends to run straight because its center of gravity rises with any turn out of plane. However, measurements show that the center of gravity of all bicycles falls as the front wheel is turned out of plane. Note that actually we are not looking for a force that prevents the front wheel from turning relative to the road. We merely want a force that would prevent the rest of the bicycle from making too large an angle with the front wheel.

The castering forces do just that! As we know, casters are those gadgets that make furniture and shopping carts roll easily. The principle behind them is very simple. A wheel carrying a heavy load will roll easily in the direction in which it is pointing. It will jam up, however, if you try to slide it sideways. The cure? Simply set the wheel axis slightly back from the swivel axis. The large force of sliding friction, perpendicular to the plane of the wheel, will soon align it with the direction of motion. A similar phenomenon occurs with a bicycle. The rider and the rest of the bicycle swivel behind the front wheel which defines the direction of motion. To the bicyclist, however, it will appear that it is the front wheel that exhibits the self-centering action.

15. Surprisingly, no! To say that the walls or columns of an extremely tall building are exactly vertical means that they are perpendicular to the plane tangential to the earth at the point of contact. This implies that each column points along a radial direction. Hence we can picture

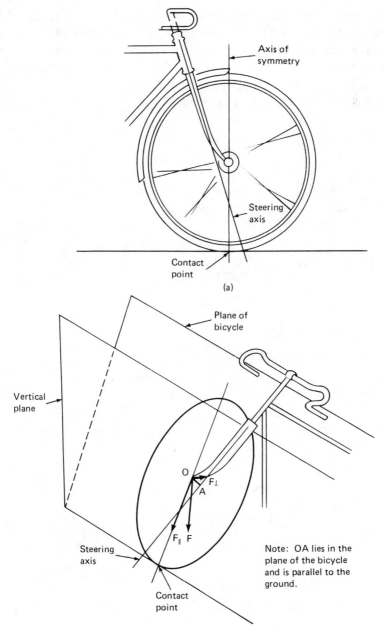

Axis of symmetry

Steering axis

Contact point

(a)

Plane of bicycle

Vertical plane

Steering axis

Contact point

O

$F_\perp$

A

$F_\parallel$ F

Note: OA lies in the plane of the bicycle and is parallel to the ground.

(b)

imaginary straight lines extending from each column toward the center of the earth, and eventually meeting there, which tells us that the columns cannot be parallel. This factor is a consideration only in the case of exceptionally tall or wide buildings, but engineers have been known to overlook it, thinking that the measurements for a small building can be transferred precisely to a large one.

3. Gases

1. Blow hard and suddenly, parallel to the table top, and the dime will hop in! As the air is blown rapidly over the top of the dime, Bernoulli's principle tells us that the pressure is lowered there. As a result, the pressure differential between the top and bottom of the dime raises it off the table, and allows it to be blown into the cup.

2. Big sails are just as strong as small sails. This contradicts the general principle that structures get weaker with increasing size. In this case the load (i.e., wind) is proportional to the area of the sail, so it increases as the square of the linear dimension rather than the cube as is usually the case when the load is in the form of weight. Hence the load per unit area is independent of the area of the sail. For the same reason big umbrellas can resist wind just as easily as small umbrellas.

3. Smokestacks are built for the purpose of enclosing a column of hot gases, thereby producing a draft. Due to their high temperature the gaseous products of combustion have a lower density than the surrounding air. Hence the weight of the hot gases in the smokestack is less than the weight of an equally high column of air, producing pressure (i.e., weight per unit area) differences that drive the hot gases up the smokestack. The difference in pressures is equal to the height of the smokestack multiplied by the difference in densities of hot gas and air, and this is how we calculate the draft. We see that a tall smokestack will produce a greater draft, thus removing the products of combustion from the furnace at a faster rate.

4. If you studied mechanics in school, your first reaction would be probably to answer this question negatively. After all, by Newton's third law action equals reaction, and the forces would therefore cancel out, as would happen when trying to lift oneself by one's bootstraps.

However, this answer is wrong! We can simulate a sailboat by using a glider on an airtrack. The wind can be created by a small battery-powered propeller, mounted on the glider so that the air is blown into the sail. If the sail is properly designed, the boat will easily sail on the airtrack. The reason is that not all of the wind generated by the fan is caught by the sail. Hence there is a net resultant force on the glider, which propels it along the airtrack.

5. According to Newton's third law forces always occur in pairs. Therefore, if the rest of the liquid exerts an upward buoyant force on the separated volume, the latter must exert a downward force, equal to its weight, on the rest of the liquid. Thus pressure is exerted on the underlying layers. A similar reasoning can, of course, be applied to air.

6. The smaller soap bubble will blow up the larger one, collapsing in the process. In the case of two balloons, the larger one will force the air into the smaller until their sizes become equal.

The basis for this paradoxical behavior of soap bubbles is that the pressure inside a bubble decreases with increasing size. The reasoning, which is not meant to be rigorous, is as follows. Consider a soap bubble of radius R (see question diagram). If the excess pressure inside the bubble is P, then the force tending to burst the bubble from within is P times the area of the bubble—$4\pi R^2$. On the other hand, the surface tension T which tends to compress the bubble acts around the circumference $2\pi R$, exerting a total force $2\pi RT$. Equating these two forces we have

$$4\pi R^2 P = 2\pi RT$$

giving

$$P \sim 1/R$$

i.e., the pressure is inversely proportional to the radius.

7. When a parachute is falling, the air passes over the outer edge of the canvas, creating turbulent vortices. These are shed in turns from one side or the other. Each vortex is a region of reduced air pressure. Hence the parachute experiences lower pressure first on one side, then on the other, and may start swinging sideways by as much as 60°. The hole in the center is designed to permit some of the incident air to continue along the central axis of the parachute, and break up the vortices that form on the top side. This in turn reduces the swinging which could be dangerous during the landing.

8. An iceboat can only move in the direction of its runners, just as a sailboat only moves in the direction of its keel. This fact gives both types of boats stability to the sideways push of the wind. If an iceboat could move faster than the wind, the wind velocity relative to the boat would have a component pointing backward (see diagram). One could then position the sail so the force *F* acting on the sail would push the iceboat forward. Hence, it is possible for an iceboat to move faster than the wind, and still be pushed forward by it. In fact, iceboats can move up to 2–3 times faster than the wind.

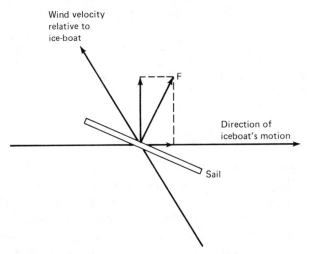

9. The problem becomes simple to solve experimentally if we recall that soap films extended between fixed boundaries take up the minimum surface area.

We construct a system composed of two parallel transparent plates joined by four pins at the corners of a square of side a drawn in the plane of one of the plates. The soap film formed after withdrawing this system from a bath of soap solution will be perpendicular to the plates, and its area will be minimized (see diagram). Since the width of the film is constant, its length will also be minimized. In fact, the length is 2.73a.

We can also solve the problem theoretically by invoking the property of planes of soap film that only three planes can intersect along a straight line. The angle between adjacent films is always 120° so all internal angles in the illustration must be 120°. Any other configuration will not minimize the area.

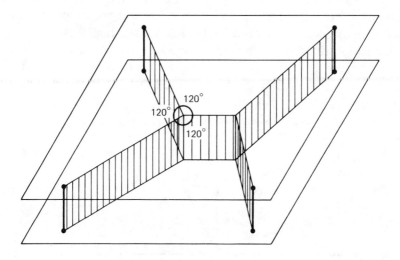

10. Sailmakers cut the canvas so that the sail is shaped like an "airfoil" when the wind puffs it out. Similarly, as in the case of an airplane wing, due to the Bernoulli effect wind passes faster along the outer curved surface of the sail, producing a reduction in pressure. The resulting pressure differences on the two sides of the sail greatly increase the speeds that can be attained by the boat. To increase the Bernoulli effect a smaller sail, called a jib, is rigged in front of the main sail. When the jib is properly rigged it channels and increases the speed of air flow over the main sail (see diagram).

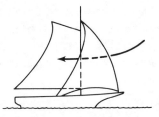

11. Large raindrops fall faster. A falling drop is subject to two opposing forces: the force of gravity mg, and the air resistance. The latter is proportional to the drop's cross section and increases with its velocity. Thus initially the retarding effect of air resistance is very small, and the drop keeps falling faster due to the constant force of gravity. As the speed of the drop increases, so does the air resistance, until the speed is so large that the force of air resistance is equal and opposite to the force of gravity. At that point the net force becomes zero, and the drop starts falling at a uniform speed called the terminal speed.

If we increase the size of the drop, the force of gravity grows in proportion to the drop's volume, i.e., as the radius cubed. On the other hand, the air resistance builds up as the cross-sectional area of the drop, i.e., as the radius squared. Therefore, as the drop's radius increases, the force of gravity increases faster than the air resistance. Consequently, the drop can reach a greater terminal speed before the air resistance catches up with it.

12. Let us examine the diagram included with the question. It shows cross sections at two points on opposite sides of the smoke ring. In the upper section the smoke particles form a vortex rotating counterclockwise as one looks into the picture; in the lower section the rotation is clockwise. The rotation of the smoke around one vortex extends to the other vortex and influences its motion. Specifically, the rotation of the smoke in the top vortex causes the smoke in the bottom vortex to move to the right; as a result the bottom vortex is swept along to the right. In exactly the same manner the rotation of the smoke particles in the bottom vortex causes the top vortex to be also swept along to the right. Under their mutual influence both vortices move to the right with the same velocity. This argument, when repeated for any pair of opposite cross sections, shows why the vortex ring as a whole moves to the right. Note how strange the behavior of the vortices appears when we contrast it with Newton's second law: force determines acceleration. In the case of vortices the action of one vortex on another does not determine the acceleration but the velocity!

13. Two coaxial vortex rings moving in the same direction attract each other, just like two electric current loops of the same sense. The vortices around one ring sweep the vortices around the other ring closer to each other, as can be easily seen from the diagram. As a result, the trailing ring is caused to accelerate, and the leading ring to decelerate. The faster ring shrinks (see puzzle 17, Chapter 4) and the slower one expands, enabling the faster ring to pass through it. Experiments show that such a passage occurs more easily if the rear vortex ring initially has a much greater speed than the leading one. However, multiple passages of vortex rings are extremely unlikely even under the best of circumstances.

14. The fallacy lies in the fact that molecules don't accelerate freely; they bump into each other and change directions. Note also that if the molecules at the bottom had greater speed than those at the top, they would be hotter. This contradicts the observation that the gas is usually a little hotter at the ceiling, because that is where the faster molecules end up. The extra pressure at the bottom of a box is actually due directly to the extra density of molecules near the bottom, which translates itself into a higher rate of collisions of molecules with the bottom surface.

4. Liquids

1. The key fact is that water does not wet the paraffin. It forms a thin film which bulges through the holes of the sieve; it is this film that keeps the water from dripping through. What's even more surprising the waxed sieve will float! This explains the tarring of barrels and boats.

2. Imagine that the tea in the cup can be mentally subdivided into a large number of tiny parcels of liquid. When the tea is stirred, each parcel is constrained to move around in a circle about the axis of rotation that passes through the center of the cup. To keep it going in a circle, each parcel must be acted on by a centripetal force $mr\omega^2$, where ω is the angular velocity of rotation and r is the distance from the center of the cup. We see that the parcels of liquid that are farther away from the center are subject to a greater centripetal force. To create such a force the pressure in the liquid must increase as one moves away from the center. However, in the bottom layers the friction from the bottom of the cup prevents the parcels of liquid from circling as fast as they do higher up. As a result the pressure difference in the bottom layer is greater than is necessary for the slower parcels of liquid to move in a circle, thus forcing them to move toward the center, dragging the tea leaves with them. To replace these portions of liquid, there is a downward movement around the perimeter of the cup, and also a rising column of liquid in the center of the cup (see diagram).

3. The submerged weight is subject to a buoyant force equal to the weight of the water in the volume displaced by the weight. Let us call this weight of water w. One might be tempted to say that to restore equilibrium a weight w should be added to the pan with the stand. However, according to Newton's third law, the force with which the

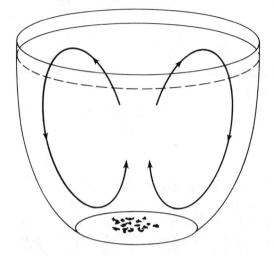

water in the container acts on the submerged weight is exactly equal to the force with which the weight acts on the water in the opposite direction. Hence, as the weight of the pan with the stand decreases, the weight of the pan with the container increases. Therefore, to restore balance a weight equal to $2w$ must be placed on the pan with the stand.

4. There is a fundamental difference between sea waves and waves in a stream. In the sea the wave *form* travels across the surface, but the water itself remains in the same place. Water particles generally move in stationary vertical circles. However, in a stream the wave form remains at the same spot with new water constantly flowing through it. This can be observed in every little stream and every river where the water flows over stones or around posts and piers.

5. Won ton.

6. Surprisingly, the answer is 10 minutes rather than 5 minutes as some people might be tempted to say. The reason is that the rate at which the water is running out is not constant. After the first glass is filled, the second will take longer to fill, because there will be less water in the container, thus exerting less pressure on the bottom. The speed with which the water is leaving the container is given by $v = \sqrt{2gh}$, where g is the acceleration of gravity and h is the height of the column of water above the orifice.

7. The velocity of the boat is the *resultant* of motion rather than its *component,* as erroneously stated in the original "solution." Consequently, the velocity v at which the rope is being pulled in is a component. Since any vector in a plane can be decomposed into two mutually perpendicular components, we get the diagram shown below. Thus v_{boat} = $v/cos\alpha$.

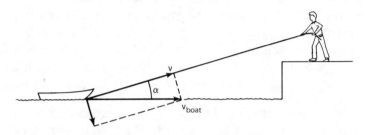

8. It is impossible to get the cork to float in the center with the setup described. The cork will always float off center and stick to the side of the glass. The reason is that when the glass is only partially full, the surface area is decreased when the object moves toward the edge. As a result, surface tension works to move the object toward the edge and hold it there.

The solution is to fill the glass above the brim. Then the cork will float in the center every time. In this case moving an object toward the edge stretches portions of the water's surface. Surface tension acts in opposition to this stretching, moving the object back toward the center of the glass (see diagram).

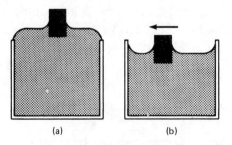

(a)   (b)

9. The matches *are* pulled. The detergent weakens the surface tension of the water between the matches. The surface tension on the outer sides of the matches is then greater than that between them, and pulls the matches apart.

10. Still at the bottom! The bucket, water, and cork all fall with exactly the same acceleration g in the earth's gravitational field (disregarding air resistance). Hence, since they all start falling simultaneously, they will not alter their relative positions while falling.

11. When the stone is in the boat, it is floating with it, and therefore by Archimedes' Principle it displaces a volume of water whose weight equals the weight of the stone. The volume of water displaced is greater than the volume of the stone since the stone is denser than water. However, when the stone is on the bottom of the pool, it can only displace its own volume of water. Thus the stone displaces less water after it is thrown into the pool, and the level of water in the pool must drop.

12. Fingers and other parts of the body are covered with a thin coat of body oils. Whenever an oily component is added to beer, its surface tension is altered, and the bubbles tend to collapse.

13. One reason is that women have a higher percentage of body weight in fat than men. The body weight of women is about 25% fat, while for men body weight is only 15% fat. Fat is lighter than water. Hence the overall density of the female body is less than that of the male body.

However, there is also another factor involved. Men have more volume around their shoulders than anywhere else. As a result their center of buoyancy (where the upward buoyant force may be considered to be centered) is in their lung area. Their center of gravity, on the other hand, is farther down in the pelvic area, due to the weight of the legs (see diagram). For women the center of buoyancy is in the pelvic area due to the large pelvic and thigh volume, and the center of gravity is a little higher. The relative position of the two centers gives women more stability while floating.

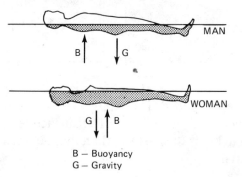

B — Buoyancy
G — Gravity

14. Paradoxically, bubbles tend to last longer if we use a liquid that has a lower surface tension. For example, plain water has a greater surface tension than water with dissolved detergent or soap, and we know that it is impossible to make long-lasting bubbles using plain water. It is easy to see why. A bubble is a glob of air trapped on all sides by a thin layer of water. If the surface tension is strong, as is the case with pure water, the outer layer of water pulls inward quite forcefully and squeezes the air inside. The compressed air then tends to burst through the film at any weak point, breaking the bubble. But if the surface tension is weak, the surface film can be more easily extended by the air. There is less tendency for the air to break through, and so the bubble tends to last longer.

15. When you pour in the water, filling the first loop, some will fall to the bottom of the second loop, and so an air trap will form at the top of the first loop. If you keep pouring water, a few air traps may form at the tops of the loops, until the pressure of the water column beneath the funnel becomes insufficient to eliminate the air traps. Past that point, no more water will enter the hose. It is also clear that none will come out the other end.

16. Use one of the additional hollow stirrers to blow air into glass A at any point where A and B glasses meet. Some of the air will enter the space between glass A and glass B, and will start bubbling to the top (bottom) of glass A, exerting enough pressure to force the water out of glass A, down the sides of glass B, and into glass C. This can be continued until all of the water has been transferred from glass A to glass C.

17. Clearly, the velocity of a vortex ring as a whole must mainly arise from the velocity the fluid rotating around one vortex has at the position of the other vortex that it sweeps along (see puzzle 13 Chapter 3). However, the velocity of rotation of each portion of fluid surrounding a vortex is inversely proportional to its distance from the axis of rotation, $v = K/r$, where K is a constant. The two diametrically opposite vortices are separated by a distance equal to the diameter D of the ring. Thus the rotation of the fluid surrounding one vortex causes the portion of the fluid at the position of the other vortex to have a velocity K/D, which is also essentially the velocity of the ring as a whole. This shows that if the diameter of a vortex ring is increased, its velocity should decrease in inverse proportion. In other words, the bigger the ring the more slowly it should move through the fluid.

We can look at the same situation from the energy point of view. Any vortex axis has associated with it a certain kinetic energy per unit of length by virtue of the motion of the rotating fluid. If now the vortex axis is bent into a circle to form a vortex ring, the energy stored in the ring will be seen to be proportional to its circumference or its diameter D. Getting back to our salt water ring, the downward force of gravity clearly does work to increase its energy. The vortex increases its diameter to store more energy, and this makes it slow down.

There is a simple physical argument that helps us understand why a vortex ring should expand (see diagram). If a force is applied at right angles to the plane of the ring, it pushes the axes of the two opposite vortices (shown in cross section) into regions where the fluid rotates in such a way that the vortex axes are driven outward. This shows why the diameter of the ring increases under these conditions.

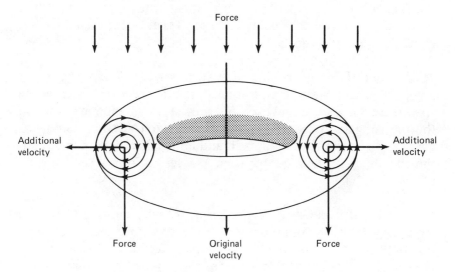

18. In the discussion below we assume that the liquid is ideal, i.e., non-viscous and incompressible. Then there is no dissipation of energy, and we can write down the Bernoulli equation

$$p + \tfrac{1}{2}\varrho\, v^2 + \varrho\, gh = \text{constant} \qquad (4-1)$$

This expresses the conservation of energy along the tube.

Next we assume that the liquid is siphoned from a container whose cross-sectional area is large compared to that of the siphon tube. This makes sure that the level of the liquid in the container goes down very slowly, effectively at zero velocity, as the liquid moves up into the siphon tube.

If the tube is of uniform bore, the velocity of flow v will be the same throughout the tube. This follows from the conservation of mass: provided there are no leaks, the mass of the liquid crossing each section of the tube per unit time must be the same, $\varrho \, Av = $ constant. Thus, since in our case ϱ and A are constant, so is v.

Now let us apply Equation (4 − 1) to point B (see diagram). At the surface of the liquid in the container Equation (4 − 1) reduces to $p_o + \varrho gh = $ constant, since $v = 0$. At the same level inside the siphon tube Equation (4 − 1) is $p_B + \frac{1}{2} \, \varrho v^2 + \varrho \, gh = $ constant. Equating these two expressions, we get

$$p_o = p_B + \tfrac{1}{2} \, \varrho \, v^2 \qquad (4-2)$$

At point F Equation (4 − 1) is $p_F + \frac{1}{2} \, \varrho \, v^2 = $ constant. Equating this with the expression at B we get

$$p_B + \tfrac{1}{2} \, \varrho \, v^2 + \varrho \, gh = p_F + \tfrac{1}{2} \, \varrho \, v^2 \qquad (4-3)$$

But at F the tube is open to the atmospheric pressure, so $p_F = p_o$. Combining (4 − 2) and (4 − 3), we get $v^2 = 2gh$, i.e., the liquid emerges at F with just the speed it would acquire by falling through a height h.

We might note that the pressures at points inside the siphon tube are all less than the atmospheric pressure, with the exception of point F. For example at B, by (4 − 2) $p_B = P_o - \frac{1}{2} \, \varrho v^2$. In general, the pressure head, i.e., the pressure above p_o, for any point is $\varrho \, gh - \frac{1}{2} \, \varrho \, v^2$. This compares with $\varrho \, gh$, the value used in arguments that disregard the motion of the liquid.

We can now easily see what maintains the flow of liquid through the siphon tube. The pressure in the container on the horizontal level of A is $p_o + \varrho gh$, since the liquid is moving down so slowly that it may be considered at rest. On the other hand, the pressure at A just inside the entrance to the tube is $p_o + \varrho \, gh - \frac{1}{2} \, \varrho \, v^2$, just as discussed in the preceding paragraph. Thus the pressure drops by $(p_o + \varrho gh) - (p_o + \varrho gh - \frac{1}{2} \varrho v^2) = \frac{1}{2} \varrho v^2$ across the entrance to the tube, and this is precisely what makes the siphon work.

It should be noted that static theories cannot explain this pressure drop since it arises through the motion of the liquid.

Moreover, the pressure drop does not depend on the atmospheric pressure p_o, for the latter cancels out in the subtraction. Thirdly, the speed at which the liquid goes through the tube is given by $v = (2gh)^{1/2}$. Thus the siphon will stop operating when the level difference h is zero.

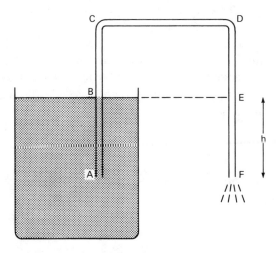

19. This question *can* be answered by a static theory. Suppose that by suction we bring the liquid down to point F (see diagram in preceding answer). Upon opening the tube at F, the pressure there is momentarily greater than the atmospheric pressure because of the added hydrostatic pressure of the column of liquid above F. The liquid then starts flowing out of the tube, increasing its speed until the pressure at F becomes equal to p_o. At this point the speed is $(2gh)^{1/2}$, and it gradually decreases as the level difference h becomes smaller.

20. The problem is that most liquids contain dissolved air, foreign matter, and tiny air bubbles. If the pressure at any point falls below the vapor pressure of the liquid, the air bubbles will rapidly grow, and break the continuity of the liquid column thus making the operation of the siphon impossible. The pressure in the liquid drops as a result of viscous friction and the loss of head due to elevation. The role of the atmospheric pressure is to compress the liquid column and make it harder for it to break. However, if there is no dissolved gas in the liquid or on the walls of the tube, then the atmospheric pressure will play hardly any role, and the uptake heights reached will be much greater than the usual 32 feet (9.6 meter) in the case of water. Dissolved gas can be removed by boiling the liquid several times or subjecting it to high pressures.

5. Living World

1. You are much more likely to snap your ankles. During impact they have to resist a compressive force equal to almost the entire mass of your body multiplied by the deceleration, i.e., by how fast you reduce your impact speed to zero. The neck must only withstand a force equal to the mass of the head times the deceleration. Another reason why the ankles are so vulnerable is that each of the two tibia a little above the ankle has the cross-sectional area of only 0.5 in^2. Hence the force per unit area at the ankle may become very large. It may even exceed the compressive strength of bone. To minimize a chance of fracture, one must reduce the speed at impact as slowly as possible. Parachutists are trained to make their first contact with the ground on their toes, and use the bending of the ankle joint to begin the process of deceleration. The knees then bend and the body is turned to the side so that the parachutist falls on the leg, then thigh, and side of chest.

2. Pigs have no sweat glands so they cannot cool themselves by the evaporation of perspiration. They can lose heat though by contact with something cool or by covering their skin with moisture, and mud is certainly perfect in both of these regards.

3. The African elephant is the biggest land mammal on earth. Hence he has less surface area compared to his weight than smaller animals. But living in a hot climate, the elephant needs a large surface area to dispose of extra body heat through radiation and evaporation. His large ears serve this purpose admirably, in addition helping to intimidate any potential aggressors.

4. A man coming out of a bath carries with him a film of water of about one fiftieth of an inch thickness, which weighs roughly a pound. A wet mouse has to carry about its own weight of water. A wet fly has to lift many times its own weight, and once wetted by water is in great danger of remaining so until it drowns. The reason for these differences is in the surface-to-volume ratio which is very large for tiny insects, and very small for large animals.

5. Small animals have a bigger surface-to-volume ratio than large animals. The surface area is proportional to the square of the linear dimension of an animal, whereas the volume varies as the cube of the linear dimension. Hence their ratio varies as the inverse of the linear dimension, which is larger for small animals. Now the heat loss in animals is mainly through the surface, so it varies like the surface area. However, the production of heat goes on in every cell of the body so it is proportional to the volume. Thus we see that small animals lose proportionally a lot more heat compared to how much they produce, and have to make up for it by consuming great quantities of food.

6. An adult cat's tail would have to be proportionately much thicker than a kitten's tail to make stable equilibrium possible in a vertical position. For the same reason, old and large trees appear rather squat or stunted compared to the slender proportions of young trees. In both cases, the forces that make a tail or a tree bend under its own weight vary roughly as a cube of the linear dimensions, while the resistance to bending increases only as fast as the cross section or a square of the linear dimensions. Hence, if you take two columns of identical proportions, the longer one will be more likely to buckle, and so will have to be made much thicker to achieve stability.

7. The upper arm is exactly at the level of the heart. Hence the blood pressure in the upper arm artery is practically equal to the pressure exerted by the left ventricle during its contraction. If the blood pressure were measured around, say the ankle, the reading obtained would be a sum of the pressure due to the contraction of the ventricle and the hydrostatic pressure of the column of blood extending from the ankle to the heart. The latter figure would vary depending on the person's height, making it impossible to compare blood pressure readings with their ideal values.

8. The illustration contains tracings made from a film at roughly 1/20-second intervals showing eight positions of a cat during its descent. There are no external torques acting on the cat so its net angular momentum must be constant throughout the free fall. In fact it must be zero if the cat was simply dropped without being imparted any net rotation. Let us now think of the cat as consisting of two halves—the front half and the rear half. As we can see in the illustration, the front half is righted first. The cat starts out by drawing in its front paws. This diminishes the moment of inertia of the front half about the body axis. Simultaneously, the hind limbs are extended, thus increasing the moment of inertia of the rear half. The cat then rotates its front half. To make the total angular momentum zero, the rear half must rotate in the opposite direction. However, since the rear half's moment of inertia is larger, it won't rotate as far as the front half. Thus the cat accomplishes a net rotation in the direction that the front half rotated.

Once the front half is righted, the haunches are then swung around. In contrast to the first stage, now the forelimbs are extended so that the moment of inertia of the front half is increased. This keeps it from changing its orientation too much. The hind limbs are drawn in, and the rear half is rotated in the same direction that the front half rotated during the first stage of the process. Now the front half rotates somewhat in the opposite direction but not as much as the rear. Thus an additional net rotation is achieved, and the cat lands the right way up.

The second stage is usually accompanied by a vigorous rotation of the tail, but even tailless cats have the ability to right themselves before landing.

9. Consider a tightrope walker initially holding his balancing pole horizontally. Suppose that he loses his balance and starts to fall over. As soon as he becomes aware of this, he can restore his balance by swinging the pole downwards—paradoxically, in the *same* direction in which he is falling. The reason is that there is no external torque, i.e., a turning moment, on the pole. The weight of the pole may be considered to act on its center of gravity, and thus cannot rotate it. As a result the total angular momentum of the system composed of the pole and the

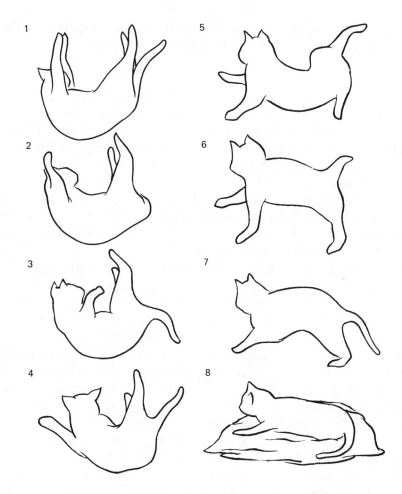

tightrope walker must remain constant—in this case zero. By swinging the pole downward the walker generates, say, a clockwise angular momentum in the pole, and therefore an equal and opposite angular momentum must be generated in the walker so that their sum will add up to zero. The counterclockwise angular momentum of the walker means that he is being turned in the direction opposite to that of the fall, and thus his balance is restored.

By slightly overcompensating, the pole can also be brought back to its initial horizontal position. Obviously, the longer and heavier toward the ends the pole is, the greater its moment of inertia and the angular momentum generated in it, thus making the balancing angular momentum greater and the balance easier to maintain.

10. The skater provided the extra energy by doing the work of pulling in her arms. At first that doesn't seem to make any sense, for if we hold an object out and pull it in, we don't do any work. But that is the case if we are not rotating. When we *are* rotating, the object has to be pulled in against a centrifugal force. The work done against the centrifugal force is exactly equal to the difference in rotational energy.

11. Yes! An unofficial speed record over a distance of 1 kilometer (0.62 mile) from a flying start is 127.25 mph. A bicyclist can go more than 100 mph once his air path is broken by a car or a train. It is interesting that a sprinter going around 25 mph could profit very little from following a car since at such low speeds air resistance is negligible.

12. The power P developed by an animal is proportional to the cross-sectional area of its muscles. Thus $P \sim L^2$, where L is the animal's linear size. In running on level ground, the animal must primarily overcome the force of air resistance which is proportional to the square of the speed and the animal's cross-sectional area. Therefore the power spent in overcoming air resistance is $P = Fv \sim v^2 L^2 v$. Setting these two expressions equal to each other, we obtain $L^2 \sim L^2 v^3$, so $v \sim L^0$. This means that the running speed is independent of the animal's size.

In running uphill the speeds are slower, so we can neglect the power necessary to overcome air resistance as compared to the power needed to do work against the force of gravity $mg \sim L^3$. Then $P = Fv = mgv \sim L^3 v$. Comparing this with the generated power which is again proportional to L^2, we get $L^3 v \sim L^2$, i.e., $v \sim 1/L$. It follows that smaller animals can run uphill faster.

13. Yes. The reaction to each bird pushing down on the air below it creates an updraft around it. If other birds crowd in close, they can use those updrafts to keep themselves aloft. In a vee formation all the birds, with the exception of the lead bird, share approximately the same amount of updraft since each bird flies forward into the updraft of the bird in front. Calculations reveal that a group of 25 birds can fly in formation some 70% farther than one bird alone.

14. A few numbers will show why branching patterns are so wide-spread in the living world. Set the distance between adjacent dots equal to one unit. Then the total length of all paths, each path starting at the center and terminating at a particular dot (or leaf), is 90 units in (a) and 233.1 units in (b). The average length of a path is 3.67 units in (a) and 3.37 units in (b). Clearly, the branching pattern in (a) has a much shorter total length than the explosive pattern in (b), at the expense of only a little longer average path length. Thus, tree branches, blood vessels, rivers, and even subway routes are all examples of branching patterns.

15. This observation was apparently first recorded by Leonardo da Vinci, but it wasn't until about 1920 that its explanation was found by Cecil D. Murray and also Wilhelm Roux. They both invoked the principle of least effort. Suppose that fluid travels in the main branch from point A to point D (see diagram). In addition suppose that some fluid also reaches point P through a side branch. If fluid flows directly from B to P, it travels within a thin branch with its large frictional loss over a long distance. The work of the fluid traversing that route is $F_{BP} \times BP$, where F_{BP} is the force required to drive fluid through the narrow side branch. Alternatively, the fluid could proceed from B to C in the main branch with its small frictional loss and then cross over to point P. In that alternate route the total work is $F_{BC} \times BC + F_{CP} \times CP$.

Which of those two routes involves less work? If the main branch is very large in comparison with the side branch, fluid will flow with less effort if it travels primarily in the large branch and minimizes its journey in the small, that is, if the small branch splits off from the main branch at close to a 90° angle. On the other hand, if the main and side branches are close to the same size, fluid will, with no increase in effort, switch over to the side branch. As a result, large branches will strike off from the main branches at angles considerably narrower than 90°. Gravity makes these theoretical angles even more narrow since narrow forks of trees make the limbs easier to support.

16. Leonardo da Vinci's rule implies that when a trunk of diameter d_o splits into two branches of diameters d_1 and d_2, $d_o^2 = d_1^2 + d_2^2$, i.e., their cross-sectional areas are equal. However, what we must consider is the resistance to flow rather than the capacity for flow. Since the

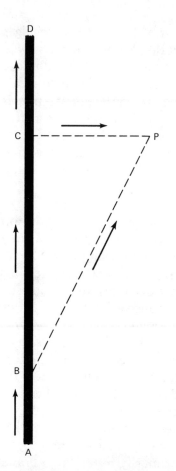

branches offer more resistance to flow than the trunk, their cross-sectional area must be increased if they are to transport the same amount of fluid. The exponent found empirically is about 2.5, that is $d_0^{2.5} = d_1^{2.5} + d_2^{2.5}$, not 3 as might be expected. The reason is that much thicker branches would be too difficult to support against gravity.

6. Sound

1. When a sound coming from a speaker's mouth encounters a wall or a ceiling, part of it is reflected and the rest is absorbed. Generally speaking, higher pitched tones are absorbed to a greater extent than lower pitched tones. Hence bass and tenor notes are reflected a greater number of times, and keep flying through the room longer before they are finally absorbed. As a result a male speaker needs to expend less energy to fill the room with his voice. However, he must contend with a longer reverberation period than a woman speaker, and so must speak more slowly. Otherwise the beginning of his next word will be heard simultaneously with the end of his last word, which may still be flying around the room.

2. The velocity of sound is given by the formula

$$v = \sqrt{\text{elasticity/density}}$$

thus the velocity of sound becomes greater as elasticity increases and as density decreases. Because solids and liquids are so much denser than gases, it might be expected that the speed of sound in these media would be slower than in a gas such as air. But in fact the elasticity of solids and liquids is, in general, very much higher than the elasticity of gases, outweighing the density factor and resulting in higher velocities in solids and liquids. However, lead has a very low elasticity, i.e., it does not spring back to its original shape when struck, like steel. Hence, it has a low sound velocity, and transmits sound poorly, which makes it a poor material for a bell. Rubber is another exception also because of its extreme sponginess, and peculiar chemical structure.

3. Low-pitched sounds can be heard for greater distances than high-pitched sounds. It is not hard to see why. As sound waves are transmitted, some of the energy that created the wave is transformed into heat energy. In sound waves with high frequencies—waves that vibrate more often in one second—the sound energy is transformed into heat more rapidly than in waves with lower frequencies.

Ships at sea need plenty of space to change course so as to avoid danger. Thus foghorns are always very low pitched to be sure their signals can be heard across miles of water. Low-frequency sounds can travel so far under water that someone once said that if there is ever a "shot heard 'round the world," it will be fired under water.

4. The sound is very soft if we listen to it in air. In air sounds spread out in all directions and their energy is soon used up. Even in a small room there is quite a lot of air—in an ordinary living room there are perhaps 2000 cubic feet—and the sound energy has to be spread all over this. In a piece of wood, even a long one like a floorboard, the space taken up is unlikely to be more than one cubic foot, and so the sound waves are mostly confined to a very small space. Hence, for example, a loud sound in air—such as striking the wood with a hammer—will be deafening and painful if you put your ear right to the wood.

5. When you pull your fingers, the pressure in the lubricating fluid between the finger joints decreases. This liberates some of the dissolved gases (mainly carbon dioxide) in the form of bubbles. The cracking noise we hear appears to be associated with the subsequent collapse of the bubbles. It takes a few minutes for the gas to become reabsorbed and ready for the next show.

6. Helium is lighter than air, so the speed of sound in helium is correspondingly greater. The resonant frequency of any cavity—including your mouth—is given by $f = v/\lambda$, where v is the speed of sound, and λ its wavelength, roughly equal to the size of the cavity. Thus the frequency increases with speed, giving your voice a higher pitch.

7. If a string is struck with a hard sharp hammer, the total energy of the sound produced is divided among a very large number of frequencies, namely the fundamental frequency and its multiples (so-called harmonics). Hence the share of the total energy for each frequency is very small. The fundamental tone and the few lower harmonics get but little of the total energy, its main bulk going into the many higher harmonics. As most of these are discordant both with the fundamental tone and with one another, the result is a sharp shrill sound of metallic quality like the noise we hear if we accidentally drop a key or a coin on a piano wire.

In contrast, when a piano wire is struck with a hammer covered with soft felt, the impact is prolonged so that by the time the hammer finally breaks its contact with the string, a substantial length of the string has already been set into motion. This reduces the energy which goes into the higher harmonics, and so avoids the harsh jangle of sounds and the discord which comes from the seventh and higher (ninth, eleventh, etc.) discordant harmonics.

8. Design (a) results in better acoustics. The sound of the orchestra travels to the people in the audience both directly and by way of a reflection from the ceiling. Multiple reflections are less important, although they prolong the reverberation time without which the hall sounds dead. If a listener hears the first reflected sound less than 50 milliseconds after the direct sound, the reflected sound will appear to reinforce the direct sound and the effect will be pleasing. If the delay is longer than 50 milliseconds, the listener will hear it as an echo and it will interfere with his hearing of the direct sound. The numbers in the diagram, indicating time delays in milliseconds, clearly demonstrate the superiority of Design (a).

9. No, the sound does not get an extra "push" due to being emitted from a speeding car! The speed of a wave, in general, is not affected by the motion of the source, but depends on the nature of the medium through which the wave travels. You will not measure an increase in speed. However, due to the Doppler effect, the sound of the approaching siren will have a higher pitch: Sound waves will be more crowded ahead of the moving siren and more of them will be going past you per second.

10. A hilltop location is better. Sound waves tend to be deflected upward, and so, move off from the hill rather than remaining concentrated within the bowl of the valley. We might note that the prevailing winds can also play a part. Sound waves, traveling into the wind, are deflected upward away from the ground and people, while those that move with the wind are deflected downward toward the ground. With this in mind, buildings can be planned so that noisy operations are kept downwind, away from the rest of the plant.

11. The microphone is picking up the sound of the person's voice as it comes out of the loudspeakers and feeding it back over and over again into the microphone. Hence, if a microphone is used primarily for speaking rather than music, it is a good idea to employ a so-called unidirectional microphone. The latter, aimed at the person's mouth, does not pick up as much feedback from the loudspeakers, and so the unwanted noise is kept to a minimum.

12. The frequency of a sound emitted by a resonating cavity is given by $f = nv/2L$, where n is an integer $1,2,3, \ldots$, v is the speed of sound, and L is the linear size of the cavity. Thus the natural frequency of a cavity is inversely proportional to its size. Hence a large animal can emit a high-pitched sound because it can either have a small nasal cavity to begin with or can contract a large cavity to a small size. On the other hand, a mouse could never roar since its nasal cavity is too small to produce a low-frequency sound.

13. Although the total range of hearing in a healthy young adult stretches from about 16 to 20,000 hertz, hearing is most acute in the range from 2000 to 5500 hertz. Part of the reason for the extra acuity in this part of the sound spectrum is that the resonant frequency of the external ear canal is in this area, providing an amplification of between 5 and 10 decibels to these frequencies. The difference in our sensitivity to various frequencies is quite striking. For example, the vibrational energy required for a softly played note whose frequency is near 130 hertz (about an octave below middle C on the piano) needs to be about 100 times as great as energy required to produce the same loudness in a note whose frequency is near 2100 hertz (the third C above middle C). On the other hand, the energy ratio required for equal loudness of these two notes when both are loudly played decreases to almost unity.

14. A telephone diaphragm is limited to the range of about 300 to 2400 vibrations per second. Speakers in transistor radios generally cut out all frequencies below about 250 hertz, the frequency of about middle C. However, the main frequencies of both male and female voices lie below those ranges. Typically, males communicate at the frequency of about 120 hertz, and females at about 240 hertz. Incidentally, the highest humanly possible note was reached by the phenomenal soprano Lucrezia Agujari, who lived in the 18th century. The note was C_7—2093 hertz. The striking conclusion is that no bass or tenor or even alto tones come out of a telephone or a radio speaker. Yet we hear male voices with absolute clearness!

The paradox is explained by noting that human speech consists of both the lowest fundamental tones and their harmonics, i.e., integral multiples of the fundamental frequencies. It turns out that the ear in addition to simply reproducing the frequencies present in a sound wave also produces new frequencies which are the sums and differences of those originally present. The difference frequency is usually the most prominent of these extra notes, giving the listener a bass note.

15. In order to hear an echo, the sound waves must be reflected back in your direction. Suppose that you are standing at the foot of a hill with a house on top (see diagram). If you now shout or clap your hands, you will be disappointed for you probably will not hear an echo. The reason is simple. Sound obeys the same law as light: The angle of reflection is equal to the angle of incidence. As shown in the diagram, all of the waves will be reflected upward into the air instead of back to you. It is

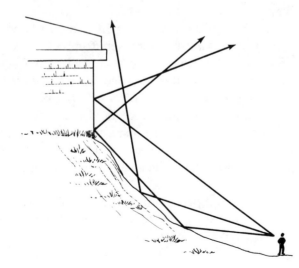

much easier to hear an echo if the sound-reflecting barrier is at the same level with you or even a bit lower, as when you are standing on a hill, and there is a building at the foot of it. The sound will come back to you by bouncing off the upward slope once or twice.

16. Western music is based on scales defined by certain frequency ratios between successive notes. In the so-called natural or ideal system, going all the way back to Pythagoras, the ratios within an octave are

C	D	E	F	G	A	B	C
1.000	1.125	1.250	1.333	1.500	1.667	1.875	2.000
24/24	27/24	30/24	32/24	36/24	40/24	45/24	48/24

This scale can be extended upward into the next octave by simply doubling all the numbers, and downward an octave by halving them. A piano tuner could adjust all the white keys on a piano to this sequence of pitches, and you could play many different kinds of simple music on it. Suppose, however, you decide to play a simple melody which originally began on C, in a new way by starting it on the next higher note of the scale, i.e., in the key of D. The result will be odd; it certainly will not sound like the original tune. And the discrepancy will be greater the farther away we move from the original key of C. For this, and other reasons, a new so-called even-tempered system was introduced over 250 years ago, in which you can play any melody equally well in every key.

In the even-tempered scale the octave is divided into 12 equal semitone intervals, so that any two successive semitones have the same frequency ratio. Since each note has to vibrate at twice the frequency of the same note an octave below, the semitone ratio from note to note is taken as the 12th root of 2, namely, 1.05946. This gives a continuous geometric progression throughout the keyboard. The following table gives the even-tempered frequency ratios, along with the ideal ratios, and the actual frequencies to which the octave above middle C is normally tuned (also see diagram for the location of the notes on the keyboard).

Note that in the even-tempered tuning the ideal C scale with which we started is approximated by the following set of intervals between successive notes:

NOTE	EVEN-TEMPERED SCALE		IDEAL SCALE
	RATIO	FREQUENCY	RATIO
C	1.0000	261.63	1.0000
C#	1.0595	277.18	
D	1.1225	293.66	1.1250
D#	1.1892	311.13	
E	1.2600	329.63	1.2500
F	1.3348	349.23	1.3333
F#	1.4142	369.99	
G	1.4983	391.99	1.5000
G#	1.5874	415.31	
A	1.6818	440.00	1.6666
A#	1.7818	466.16	
B	1.8877	493.88	1.8750
C	2.0000	523.25	2.0000

C   D   E   F   G   A   B   C

tone tone semitone tone tone tone semitone

On a piano keyboard the tuner does not make any difference in tuning between the black and white keys—they are all arranged in a uniformly rising sequence of pitch. The two colors and shapes of keys are there only to help the player find his way by feel over the wide expanse of the keyboard. Ultimately, with the even-tempered system the sequence of pitch is not in precise agreement with the natural scale, but it does provide a sufficiently close approximation. In fact, the modern ear (since the time of Bach) has become totally accustomed to the "errors," so that this tuning sounds correct.

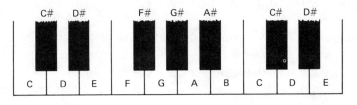

7. Heat

1. The temperature of molten lead is about 327° C (about 600° F), so the experiment is very dangerous. However, it can be done by wetting your hands first because it takes a lot of energy to heat water. The hot lead around the hand will first bring the water on the hand to boiling, then it will vaporize it. By then the hand had better be out or the consequences will be tragic.

2. The trapper who saved the travelers in Cooper's *Prairie* apparently knew a simple law of physics. Although the wind was blowing from the burning prairie into the faces of the travelers, in front of them, near the flames, a reverse air current was blowing toward the fire. Indeed, warmed up by the fire below, the air above it grows lighter and is pushed up by cooler fresh air flowing in from the travelers' side of the prairie.

3. The mercury went through the constriction during expansion due to the enormous pressure then generated. The constriction is so narrow that to make the mercury go back we have to apply reverse pressure by swinging the thermometer in an arc.

4. Some people say erroneously that the water rises to fill the space previously occupied by the oxygen that was used up by the burning candle. That cannot be the reason because the burning itself releases gases and vapors, notably carbon dioxide and water. What actually takes place involves more physics than chemistry. As the air trapped under the jar is heated by the flame, it expands and some of it may even bubble out through the submerged mouth of the jar. When the flame grows dimmer, the air cools and contracts, and then the atmospheric pressure outside the jar forces the water up into the jar.

5. Clean snow is a good reflector and so does not melt rapidly in sunlight. If the snow is dirty, more radiant energy from the sun is absorbed and the snow melts faster. It is safer for the snow to start melting gradually before the onset of spring. This prevents the sudden run-offs that often cause floods.

6. Half of the initial potential energy was converted to heat due to the internal friction in the liquid itself and the friction against the walls of the container. If there were no friction, the liquid would oscillate between the two vessels forever. This phenomenon can be approximated using nonwetting liquids such as mercury.

7. Hot air can hold more water vapor, i.e., water in gaseous form, than cool air. This is because water vapor molecules which would coalesce and liquefy upon low-speed collisions, bounce apart and remain in the gaseous state after high-speed collisions. When the air is cooled, some of the water vapor in the air condenses into tiny droplets which counteract evaporation of perspiration. Thus the extra humidity must be eliminated to make us feel comfortable.

8. The sense of touch is indeed misleading when it comes to judging the temperature. The doorknob feels colder because as a metal it conducts heat away from your finger faster than the wooden door.

9. Unfortunately not! The temperature rise inside the open refrigerator will alert the cooling system to start cooling again. But more heat will be released by the motor in the back of the refrigerator than absorbed by the cool air coming out in front. This follows from the second law of thermodynamics. As a result, the room will become even hotter.

10. The answer is, paradoxically, that the total energy of the air in the room remains the same. When the heat is added, the air in the room expands, and the only way it can do so is by leaking to the outside through the pores and cracks in the walls. The leaking air carries with it the energy added by heating.

We can easily prove that the energy content of the air in the room is independent of the temperature. The energy content of the air per unit volume is proportional to its density d and its temperature T, i.e., $E \sim Td$. However, to the extent that the air is an ideal gas, it obeys the ideal gas equation $p/(Td) = c$, where c is a constant. From this we get $d = p/(cT)$, which substituted in the first equation gives $E \sim T$ $p/(cT) \sim p$. Thus, the temperature is canceled out, and we are left with the result that the energy of the air per unit volume is proportional to its pressure but independent of the temperature.

11. No. Both the inner and outer diameter increase in size. If this were not the case, it would not have been possible to shrink a metal tire onto a wheel, as was the practice in the horse-and-buggy days.

12. When you pour hot water into a glass, the inner glass layer expands first, exerting a strong pressure on the outer layer whose expansion is delayed due to low thermal conductivity of glass. If the water is hot enough, the glass may crack. However, when you put in a metal spoon, the hot water will rapidly lose some of its heat to it. Its temperature will drop, and by then the glass has warmed up, and more hot water won't crack it.

13. There is nothing paradoxical about it because the water at the bottom of the tube remains cold. As it expands due to heat, the water becomes lighter; it does not descend to the bottom, but stays in the upper part of the tube. Moreover, water is a very poor conductor of heat, so transfer of heat by conduction to the lower part of the test tube is also difficult.

14. Both spheres expand when the heat is added. The center of gravity of the sphere on the table goes up. This means that some of the heat added goes into work against the force of gravity. Thus, the temperature rise is not as large as might be expected. On the other hand, the center of gravity of the suspended sphere goes down. The potential energy thus released goes into an additional temperature rise. We conclude that the suspended sphere will be slightly warmer than the sphere on the table.

15. During the heating season essentially no energy can be saved by turning off indoor lights. This is because the lamp itself is a very efficient heater. Most of the energy supplied to any electrical lamp is converted directly to heat (I^2R "losses"). Even the energy emitted as light, upon being absorbed by the walls, furniture, and people, is converted to heat. The heat no longer produced by an extinguished lamp must be supplied by the building's heating system in order to maintain the temperature setting of the thermostat. If a building's heating system burns fuel inefficiently (as compared to that of generation and transmission of electricity) it may well take more fuel to maintain the temperature with the light off than with it on!

One may, however, save a little *money* by turning out lights because electricity is generally the most expensive way to heat a building. Also, burned out light bulbs are not cheap to replace.

During the summer, the situation is much different. If a building is air conditioned, any heat added by lights must be removed by the air conditioner. Therefore, a neglected light wastes not only the energy delivered to it, but also the energy the cooling system used to remove the lamp's energy.

16. No. Pure water boils at 100°C, so it can heat the water in the bottle up to the temperature of 100°C. However, as soon as the temperatures become equal, no heat will flow from the boiling water to the water inside the bottle. Thus the boiling water will be unable to give it the extra heat necessary to transform it into steam, and if there is no steam, there is no boiling.

17. Yes. Under very high pressures water becomes solid, and stays in that state at temperatures way above 0°C. The American physicist Bridgman obtained what he called "Ice No. 5" by applying a pressure of 20,600 atm. This ice stayed solid at a temperature of 76°C, which is hot enough to burn your fingers.

18. Freezing mixtures employ the trick of mixing salt and ice to make a material supply the heat needed for its melting from within itself. First of all, adding salt to water lowers its freezing point because salt molecules come between the water molecules that ordinarily would join together. Thus more heat must be removed from the mixture for water molecules to slow down to the point where they can begin to fuse together. The lowering of the freezing point means that some ice present in the mixture will melt immediately. However, melting requires quite a lot of heat energy—80 calories per gram of ice, to be specific. The heat to melt each gram of ice comes from the still unmelted ice and water nearby. The salty water percolates among the granulated ice, more of which melts, removing more heat energy from nearby ice and water. Finally, the freezing mixture consists of a very cold salt solution in which the last granules of ice are melting. In microscopic terms, energy has been borrowed from the translational degrees of freedom of water to first melt the ice, that is to increase the potential energy of ice molecules, and then to get them moving. Since the original kinetic energy of water molecules is now shared among the original water molecules and those that used to be bound in ice, the kinetic energy per molecule is lower. This is the same thing as saying that the temperature of the mixture is lower.

19. When a water drop first touches a hot stove, its bottom portion vaporizes instantly. The steam blowing downward pushes the drop up out of contact with the stove, and thereafter the drop rests on a cushion of its own steam. Steam, like any gas, is a poor conductor of heat, thus preventing further rapid transfer of heat to the drop. Then a couple of minutes may pass before the drop is brought to boiling temperature.

Liquid air has a temperature of about $-196°C$. Compared with that, the hand is a pretty hot stove. Thus a drop of liquid air put on the hand assumes the spheroidal state, and if not allowed to sit in one place too long, will produce no harm. Frozen mercury at $-39°C$, in contrast, will produce a severe and painful burn.

20. The one containing hot water!

This paradoxical phenomenon was already reported by Francis Bacon in his *Novum Organum* (1620). In places that experience long winters, such as Canada and the Scandinavian countries, it has become part of everyday folklore. For example, it is believed that a car should not be washed with hot water because the latter will freeze on it faster

than cold water, or that a skating rink should be flooded with hot water because it will freeze more quickly.

First, let us point out that this is not true of pails with lids. They cool exclusively by the conduction of heat through the walls, and this process obeys Newton's law of cooling. The hot pail will first have to cool to the initial temperature of the cold pail, and from that point on it will follow the same cooling process as the cold pail. Thus, the total cooling time for the hot pail will be longer due to the initial time segment.

Pails without lids cool partly by Newtonian conduction and partly by evaporation of the contents. For wooden pails the cooling is mostly by evaporation, since wood is a poor conductor of heat. In the case of metal pails, a great deal of heat is transmitted through the walls, so now a pail with cold water will cool first. Another important factor is the temperature of the water. At high temperatures evaporation completely dominates conduction. What happens is that rapid evaporation from the surface makes the top layer cooler. Its density, therefore, becomes lower than that of the water underneath. As a result, the top layer loses its buoyancy and, by Archimedes' principle, sinks lower. The hot water underneath rises higher, and the whole convection process begins again.

Evaporation combined with convection results in a very large rate of heat loss, especially if the starting temperature is high enough. This explains why under certain conditions a hot pail will cool faster than a cold one.

We might add that the mass loss in cooling by evaporation is quite significant. For example, water cooling from 100°C will lose 16% of its mass by 0°C, and another 12% on freezing. The total mass loss is, therefore,

$$16\% + \frac{12}{100}(100 - 16) = 26\%$$

8. Electricity and Magnetism

1. When the flame is between the oppositely charged poles, the free electrons that have been shaken off the atoms by the intense heat move toward the positive pole. The flame is then left with extra positive ions and becomes positively charged. The ions are relatively heavy and don't move as easily as the electrons. Hence the flame merely tilts toward the negative pole instead of traveling toward it the way the electrons traveled toward the positive pole.

2. One may explain the purpose of keepers in a simple way by referring to the concept of the magnetic pole. Consider the circular horseshoetype magnet without a keeper (see part [a] of diagram). The free poles at the gap exert a demagnetizing effect on the magnetic domains in the iron in that the magnetic forces of these poles tend to oppose the desirable magnetic alignment of the domains. Thus, with these free poles on hand disalignment of the domains is relatively easily achieved by thermal and mechanical shocks.

If, on the other hand, a soft iron keeper is used to close the gap, then (see part [b] of diagram) there are no free poles. There is no net magnetic force at either end of the magnet and hence no force which could cause disalignment of the domains.

A note of caution is in order. The concept of the so-called free poles is only intended to convey the fact that there are net unbalanced magnetic forces at either end of the magnet.

3. Tires in contact with the road develop a negative charge. As a result, the electrons in the metal body of the truck are repelled by the tires, and leave the body area near the tires positively charged. Sparking may then occur between the truck and a nearby grounded object,

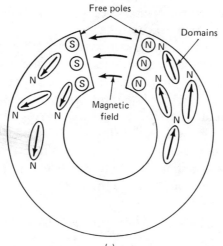

(a)

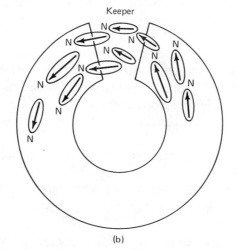

(b)

which could ignite gasoline fumes. Unfortunately, even though a chain will drain some of the electrons from the truck's body, the truck will then be left positively charged and still prone to sparking.

4. The paradox arises from trying to apply direct current analysis to an alternating current phenomenon.

With an open-circuited secondary, only a small primary current flows. This is because the primary voltage is nearly canceled by a back electromotive force (emf), according to Lenz's law. When the secondary is connected to a load, the secondary current produces a reacting magnetic field. This magnetic field tends to reduce the flux in the core and since the value of the primary back emf is dependent upon this flux, the back emf is reduced, and the primary current increases. The greater the secondary current, the greater its reacting field; the more the flux in the core is decreased, the less back emf there is, and the primary current increases that much more. In fact, the primary current will always increase so that the secondary current is equal to one-fourth its value. The power equation will then be

$$I_p U_p = (1/4 I_p)(4 U_p) = I_s U_s$$

where the subscripts p and s refer to "primary" and "secondary," respectively. We see that power out equals power in, so energy is indeed conserved.

5. Align the bar with the magnetic lines of the earth, i.e., roughly with the north-south direction. Strike it lightly a few times with a hammer. The hammering jostles the magnetic domains in the bar so they can better fall into alignment with the earth's field. To demagnetize the bar, strike it when it is in the east-west direction.

6. It turns out that there is a certain distance between the electron and the proton at which the electron feels most comfortable. For that distance, called the Bohr radius $a_0 = 0.528 \times 10^{-8}$ cm, the energy of the proton-and-electron system reaches a minimum, and the system becomes stable.

Why is there such a most comfortable distance? The reason is that the electron is acted on by two opposing forces. On the one hand it is strongly attracted to the proton by the electric force acting between them. On the other, the confinement of the electron to a small region of space implies that its position is known to within the radius of that region. However, according to Heisenberg's uncertainty principle, the smaller the uncertainty in position, the greater the uncertainty in the electron's momentum. It is as if the electron starts kicking in protest if we try to confine it to a small volume of space. Thus, in effect the electron behaves as if it doesn't want to stay that close to the proton, i.e., it acts as if it is repelled by it. As a result the electron makes a compromise between the attractive and the repulsive force, and stays at the most comfortable distance.

7. The north-seeking end of a compass needle does not point toward the geographic north pole, but rather in the direction of the magnetic north pole. The latter is located in the vicinity of northwest Greenland, about 1200 miles from the geographic north pole. As a result, in British Columbia a compass needle deviates much farther from true north than in Vermont.

8. If the electric load is largely inductive as in electric motors used to drive industrial machinery, air conditioners, home shops, etc., the current will lag the voltage. If the load predominantly includes elements of capacitance, the current will lead the voltage. In either case, the so-called power factor ($\cos\phi$, where ϕ is the angle of lead or lag) is reduced, and the users are not able to draw as much power as they would like to. Residential loads tend to be largely resistive, and have a comparatively high power factor. Industrial loads, predominantly including electric motors, are inductive in nature. In either case, if the load includes too many inductive elements, as when people use electric motors for air conditioning and home shops, banks of capacitors are needed to bring the load back into balance.

9. For simplicity, let us consider the force between a wire carrying a current and a moving negative test charge. The wire is neutral and thus can be represented as consisting of two rows of equally spaced electric charges, one row positive, and the other row negative (see part [a] of diagram).

Suppose that the negative charges (electrons) move to the left and so does the negative test charge. From the reference frame of the test charge, the positive charges in the wire move more rapidly to the right than the negative charges in the wire. However, according to the special theory of relativity an object moving at a speed v appears to a stationary object as contracted by a factor $\sqrt{1 - v^2/c^2}$. Thus, the more rapidly moving positive charges will appear to be spaced more closely together than the negative charges (see part [b] of diagram). Since the quantity of charge on each particle is independent of the particle's speed, it follows that a length of wire will appear to the negative charge as containing a greater amount of positive charge than negative. Thus, the negative charge will feel an attraction pulling it toward the wire. If the test charge is one of a row of charges in a second wire, the second wire will be attracted toward the first.

We usually call this attraction a *magnetic* force but it is nothing more than an unbalanced electric force. Being the unbalanced difference between two electric forces, the magnetic force is smaller by a factor of $(v/c)^2$. Since the electron drift velocity v is on the order of 1 cm/sec, that factor is 10^{-21}—an enormous difference. The magnetic forces can turn motor armatures and lift weights because of two factors. First, the number of conduction electrons in a piece of wire, each contributing to the magnetic force, is astronomical. A 1-centimeter length of 1-millimeter diameter copper wire carries about 6×10^{20} conduction electrons. Second, the electrostatic forces between large objects cancel almost exactly. Otherwise, they would completely overshadow the magnetic forces. In the atomic domain where the full Coulomb force between elementary particles comes into play, magnetic effects do indeed take a distant second place compared to electric interactions.

By the way, the existence of the magnetic force shows that relativistic effects can be very strong even at speeds as low as 1 cm/sec, contrary to the usual assertion that they can only be observed at speeds close to the speed of light.

10. The simple reason is that the AM radio transmitters also have vertical antennas. This means that the electric field in the electromagnetic wave broadcast by the transmitter is polarized vertically. For maximum signal power the receiving antenna should also be vertical.

11. You can feel relatively safe inside an airplane during a thunderstorm, and the same goes for an automobile.

First of all, it is very unlikely that an airplane or a car would be struck in the first place, for neither is grounded, and lightning takes the easiest path to the ground.

Secondly, a closed metal box acts as an electrostatic shield. Electrons from the lightning bolt are mutually repelled to the outer metal surface of the airplane, and arc to the ground or a cloud, leaving the occupants safe inside.

More generally, electric charges outside a hollow conductor cannot even set up a field inside the conductor or induce a charge on its inner walls (see diagram). Why? Because if there were electric lines of force in the hollow interior of the conductor, they would begin and end on the same conductor. But this is forbidden since an electric line of force may only join two points if they are at a different voltage, and a metal surface must all be at the same voltage.

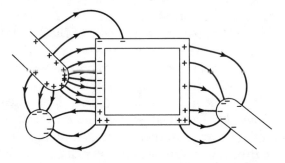

12. Yes. if a charge $+Q$ is placed inside the shield, a total charge $-Q$ will appear on the inner surface of the shield (see diagram), and a total charge $+Q$ on its outer surface. All three charges combined will establish the same electric field outside the shield as the field that would be created by the single unshielded charge $+Q$. Therefore, we cannot screen a charge in this manner.

However, it is possible to salvage the situation by grounding the metal shield, i.e., connecting it directly or indirectly to the earth by a thin wire. This will cause the charge on the outer surface of the shield to be transferred almost entirely onto the surface of the earth. This occurs because of the tremendous capacitance, or the charge storing ability, of the earth, the capacitance of a spherical conductor being proportional to its radius. Since the lines of force between the charge $+Q$ inside the shield and the induced charge $-Q$ on its inner surface cannot penetrate the metal wall of the shield, to an outside observer the charge

+ Q will be very effectively screened. We may note that the shield may be as thin as a sheet of foil or a coat of aluminum paint.

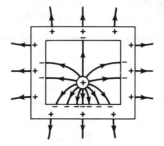

13. Yes, by making a magnetic tripole. Normally, a bar magnet can be made by stroking a bar of steel with a magnet or placing the bar momentarily inside a coil carrying a current (see diagram [a]). Imagine the unmagnetized bar as consisting of many tiny magnets pointing in random directions. The passing north pole of the magnet will clearly pull all the south poles of the tiny magnets toward itself, thus producing the magnetization indicated in the diagram. Now, suppose that we use two magnets, each stroking one half of the unmagnetized bar (see diagram [b]). Incredible as it may seem, this will produce a magnetic tripole, technically known as consequent poles. One tripole will, of course, always float above another tripole, regardless of their relative orientation. A tripole can also be made by placing a steel bar inside a coil wrapped so that the direction of wrapping is reversed at the center of the coil.

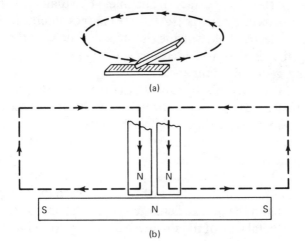

14. Paradoxically, even though parallel electron beams seem to constitute like currents, the beams will repel! The reason is not too difficult to see. The total force between moving charges consists of the electric and magnetic force. In the case of two wires carrying like currents, the charges of the positive ions and the conduction electrons neutralize each other in each wire, so the electric forces between them are precisely zero. There could be some net electric charges on the surfaces of the wires, but they may be neglected assuming voltage differences are small. As a result the only forces left in the picture are the velocity-dependent magnetic forces. Normally, they are extremely small compared to the electric forces (see puzzle g). However, for two wires we are adding magnetic forces between some 10^{20} electrons! No wonder the resultant attractive magnetic force can be easily observed.

In the case of two parallel electron beams, the attractive magnetic forces between the beams are very weak despite high electron velocities. This is due to the fact that the numbers of the electrons in the beams are minuscule as compared to the number of conduction electrons in the wires. Consequently, the repulsive electric forces between the electrons easily predominate. However, as the electrons approach the speed of light the magnetic forces increase until they become nearly equal to the electric forces. Then the net force between the electron beams approaches zero.

15. Surprisingly, it is much harder to build an effective shield against a magnetic field. However, one can weaken it considerably with a thick ferromagnetic housing. As we recall, ferromagnetic materials such as iron, nickel, cobalt, and their alloys have a domain structure. Each domain is a tiny but powerful permanent magnet. Normally, the domains are randomly oriented. In an outside magnetic field, the domains become partially aligned and the field inside the ferromagnet may become extremely strong. We represent this pictorially by showing more densely spaced magnetic lines of force inside the ferromagnetic housing (see diagram). Clearly, the magnetic field concentrates itself in the housing instead of penetrating strongly into the interior region. If the housing is thick enough, the inside field can be reduced by a factor of more than a thousand. Similarly, a ferromagnetic shield placed around a current-carrying coil weakens the magnetic field of the coil outside the shield.

This whole discussion applies also to slowly varying (up to several thousand cycles per second) magnetic fields. However, for high-frequency magnetic fields a housing made of a good conductor such as copper or aluminum is a better shield than a ferromagnetic housing. At high frequencies the shielding against magnetic fields involves the generation of counter magnetic fields by eddy currents induced in the conducting housing. Interestingly enough, the same method can be used to provide protection from electromagnetic interference.

Electromagnetic shields make use of the fact that an electromagnetic wave must simultaneously contain electric and magnetic fields in order to propagate independently through space. If we eliminate either the electric or the magnetic components of the wave, the other component is also halted. Therefore, since a shield designed to eliminate electrostatic fields can be made quite easily, this type of shield is also used to eliminate the electric-field component of an electromagnetic wave. Without the electric component, the magnetic field cannot continue to propagate and is thereby extinguished too. Thus, an enclosure consisting of a good electric conductor connected to ground through a low-resistance path will be an effective shield against electromagnetic interference.

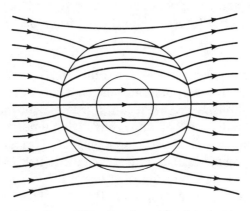

16. Imagine a magnified piece of wire placed in a magnetic field that points into the page (see diagram). Suppose a current flows in the wire toward the top of the page. There will be a sideways force $F = -qv \times B$ on the electrons constituting the current. As a result, the electrons will tend to drift toward the right. An excess of electrons on the right, and a deficit on the left will create a repulsive force on the electrons drifting toward the right. This is known as the Hall effect. The electrons will keep piling up on the right until the repulsive force becomes strong enough to counterbalance the force due to the magnetic field and there is no longer a net force on the electrons. Note, however, that the positive ions of the metal have fixed positions, and no magnetic force exists on them. But they are very much subject to the electric force due to the pileup of the electrons on the right. This electric force pulls the ions to the right, thus producing the motion of the wire as a whole. The paradox is therefore resolved by observing that the motion of the wire is caused by an electric field, not a magnetic field.

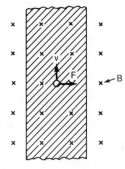

17. Suppose that initially the top-left can A is charged negatively slightly more than the top-right can B. The initial charge asymmetry is purely accidental, and may be due to the charging by cosmic rays, the earth's natural radioactivity, or even due to the residual charge left from the previous run of this experiment. Then the bottom-right can D will also be more negative than the bottom-left can C because of the wiring. As the water streams falling from the nozzles break into drops, the drops falling through can A will be positively charged, the negative charge in the stream being repelled by the can. The positive charge trapped in the drops is then carried by gravity away from the attracting can A and into the positively charged can C against the force of electrostatic repulsion. This explains why the drops tend to break up into fine spray as they approach the bottom cans. Consequently, positive charge accumulates in can C which was positive to begin with. The identical mechanism is at work on the other side, where the charge accumulation in can D is negative. A visible spark will occur when the cans are discharged.

18. Some textbooks do in fact claim that in the second experiment the capacitor plates repel. This is utter nonsense. Oppositely charged plates must attract. What some people forget is that to maintain a constant voltage difference between the plates work had to be done to remove some charge from the plates as the capacitance was reduced. Specifically, an amount of charge given by $V\Delta C$, where ΔC is the decrease in capacitance, was removed at a voltage V. So the work done was $V^2\Delta C$ plus some work W went to overcome the attraction in moving the plates farther apart. These two energy expenditures add up to a total energy loss given by $1/2\ V^2\Delta C$. Thus $W = -1/2\ V^2\Delta C = F\Delta d$, where Δd is the increase of the separation. Since ΔC is negative, F is positive. Consequently, a positive force was needed to move the plates farther apart, as it should be if the force between the plates was attractive.

19. The relatively low frequencies (540 to 1600 kilohertz, i.e., 540,000 to 1,600,000 cycles per second) used for AM (amplitude modulation) transmission correspond to wavelengths of 200 to 500 meters. Electromagnetic waves of such length are easily absorbed by large objects. This is why a pocket radio is unsatisfactory when used in a steel frame building. FM transmission, on the other hand, makes use of very high frequencies (VHF), ranging from 88 to 108 megahertz, i.e., from 88 to 108 million cycles per second. These correspond to wavelengths of about 3 meters. In fact the FM radio band is situated right in the frequency gap between television channels 6 and 7. Signals in this frequency range, including television signals, are not absorbed by large objects, such as buildings or bridges, but are reflected from them and scattered in all directions. Occasionally, both a direct and reflected signal from the same station may be received at the same time. On TV this causes "ghost images," and on FM stereo it results in distortion or noise. However, barring such events, FM reception is not affected seriously by large objects, particularly in strong signal areas.

9. Light and Vision

1. A rainbow is a part of a circle. Its center is easily seen to lie below the horizon along a straight line passing from the sun through the observer's head. The rays from the rainbow to the eye make an angle of about 42° with the axis of the rainbow (see diagram). The lower the sun descends, the more the center of the rainbow ascends, more and more of the circumference appearing above the horizon until it becomes a semicircle as the sun sets. On the other hand, when the sun is higher than 42° the rainbow disappears completely below the horizon.

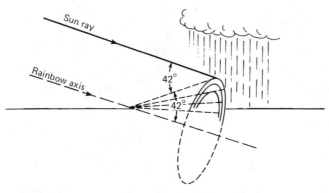

2. When a substance is heated, the outer electrons (i.e., those farthest away from the nucleus) can use the heat energy to jump to higher energy levels. When the electrons jump back down to lower energy levels, they get rid of the extra energy by radiating electromagnetic waves. If the waves are in the visible portion of the spectrum, we observe them as glow. Now, in metals the outer electrons are rather loosely attached to the nucleus and it doesn't take much energy to make them jump to higher energy levels. They also easily jump back to lower levels. On the other hand, in quartz the outer electrons are tightly bound to the nuclei, and even a temperature of 800°C is not enough to produce mass jumps to higher energy levels.

3. The contradiction is only apparent. The eye of an observer receives reflected light from the bottom of the stick, B (see diagram). But the light ray from B bends at C following the path BCD to reach the eye. To the observer the light appears to have come straight from behind C, or from a point around E. Notice that point E is higher than B, so the stick appears to be bent upward.

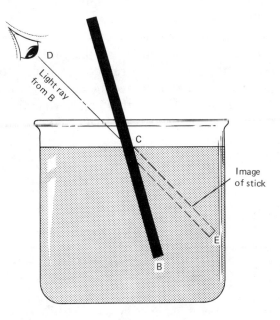

4. An image of the outside world is focused on the retina composed of a mosaic of rods and cones. Their diameters are roughly equal to the wavelength of light. If the eye is to distinguish between two small objects, their images must fall on separate elements. Thus with fewer rods and cones, the image of the outside world would be subdivided into a smaller number of pieces, and we would see less distinctly. On the other hand, the size of rods and cones can't get much smaller for then they would not respond to visible light at all. Remember that rods and cones are in a way like antennas which must always be the size of the wavelength received. Thus we see that the eyes must have a certain minimum size to be of any use at all. However, there is no need for them to be much larger. Consequently an elephant's eyes are only slightly larger than ours.

5. The paradox stems from the fact that "up" and "down" are directions whereas "left" and "right" are orientations. As such these two sets of concepts belong in two totally different categories. Up and down are global concepts, like east and west. Two observers standing at the same point will agree on the east and west directions as well as on which way is up or down no matter what direction they are facing. On the other hand, left and right are local concepts associated with the frames of reference attached to the body of each observer. As a result, two observers facing each other will disagree on which way is right or left.

Suppose that a person is standing in front of a vertical mirror aligned with the east-west direction so his left arm points west and right arm east. Then the images of the left arm and the right arm still point west and east, respectively. There is no reversal of directions here, just as there is no reversal of top and bottom. On the other hand, the person and his image are facing each other, so naturally they will disagree about the directions of left and right.

6. The pinhole monocle works because it converts the eye into a pinhole camera which has infinite depth of field. The eye itself can automatically produce the same effect. In very bright light the pupil decreases in size thus cutting out some of the aberration and improving vision. Squinting is based on the same principle of reducing the amount of poorly focused light. Moreover, one can make a pinhole using the fingers of one or both hands. This is useful while swimming when glasses may not be at hand.

7. On a clear day the sky appears blue because the atmospheric molecules scatter the shorter wavelengths of the blue and violet colors more than the red and yellow colors from the incoming beam of "white" sunshine. However, haze, fog, and cloud droplets are as large or larger than the wavelength of sunlight, and they do not show a preference for scattering any particular wavelength; all are scattered equally. Thus when the atmosphere is hazy or cloudy, the sky appears white.

8. One reason is that we evolved this way because 83% of the sunlight that reaches the earth falls in our range of perception. Moreover, the human eye is most sensitive to light in the yellow-green range which also happens to coincide with the peak of the sun's energy output.

Another explanation is found by examining the mechanism of vision. Wavelengths much shorter than those of visible light, such as X-rays or ultraviolet radiation, would carry enough energy to destroy the organic molecules in the eye. Longer wavelengths such as infrared radiation and microwaves, on the other hand, would not have enough energy to trigger changes in molecules of the retina. As a result no visual impulses would reach the brain.

9. Even though glass looks transparent, only part of the light that strikes a window pane is transmitted. A portion of it is reflected back into the room. It is this reflected light that is responsible for the mirror-like behavior of the window at night. In the daytime the small amount of reflected light is masked by the much brighter daylight.

10. Very easily, and in at least two ways, as shown in the illustrations. In (a) the mirrors should be at right angles to each other, with two edges touching. In (b) you bend a sheet of thin metal polished enough to give a mirror reflection until you obtain an undistorted image of your face.

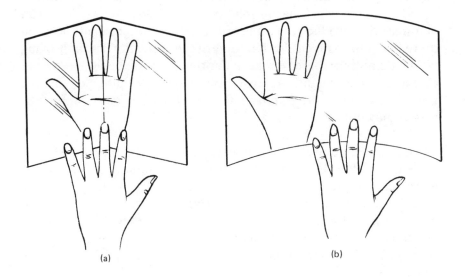

(a)   (b)

In both cases the image is true and unreversed. You can test this by winking, say, your right eye. Instead of your image's left eye winking, i.e., the eye directly opposite your right eye—the image's *right* eye winks. Thus, for the first time, you can see yourself in a mirror exactly as others see you!

11. The explanation predicts the wrong direction for the rotation of the vanes. They actually rotate in the direction toward which the *white* sides are directed *unless* the vacuum in the glass bulb is almost perfect. Then and only then is the explanation correct. Otherwise, the gas forces predominate. But even in this case the usual argument is wrong. It states that the molecules of the remaining air that bombard the warmer black side rebound with a greater speed, and so impart a greater recoil to the black side than the molecules that bombard the cooler white side. This results in the correct sense of rotation of the vanes, but does not explain, for example, why the radiometer works only at reduced pressures.

What we must realize is that the pressure on a surface is not only proportional to the momentum transferred by the individual molecular impacts, it depends as well on the rate at which the molecules are impinging on the surface. The rapidly rebounding molecules from the warmer surface are more effective in stopping other molecules from approaching and hitting the surface. This is equivalent to saying that any local increase in the temperature of the air is quickly counterbalanced by a lower air density. As a result, the pressure over most of the warm surface is the same as the pressure on the cooler side. But for molecules impinging near the edge of the warm side, onto a strip of the order of the mean free path in from the edge, the situation is different. They are held back partly by molecules rebounding from the vane and partly by molecules passing the edge of the vane from the cooler side. However, the latter are less efficient in stopping incoming molecules, so more collisions per unit area and time will occur near the edge of the vane. The pressure near the edge, then, will be greater than at the center of the warm side, and therefore also greater than the pressure on the other side, thus producing the rotation of the vanes.

This explains why the rotation is observed only at reduced pressures. The excess pressure is exerted only on a strip about one mean free path in width that runs along the edges of the vanes. The strip becomes thinner as the pressure rises. At intermediate pressures, convective currents in the unevenly heated gas are dominant and generally tend to move the vanes in the opposite sense, while at atmospheric pressure both effects are so small that the vanes do not respond.

12. The radiometer will spin, but in the direction opposite to the usual. The black surfaces, being good absorbers of light and heat radiation, are by Kirchhoff's law also good radiators. They will cool down faster than the white surfaces, and so in this case an excess air pressure will exist along the edges of the warmer white surfaces.

Interestingly enough, a radiometer placed in a refrigerator or in front of a heater will gradually cease its rotation, as both the black and white surfaces of the vanes attain the same overall temperature. A radiometer run by sunlight will not stop spinning since it can never come to equilibrium with the temperature of the incident radiation, i.e., the temperature of the sun's surface—6000°C.

13. Light, like all energy, travels in waves, just as the energy of motion travels in waves across the ocean. Ocean waves oscillate (vibrate) vertically, describing an up-and-down motion as they move toward the shore. Light waves from the sun and from ordinary light bulbs vibrate at all different angles: horizontally, vertically, and at every angle in between.

When you look down a brightly lit roadway, with the sun's glare reflecting off the hood of your car and blinding you, the light reaching your eye from the glare spots contains a large percentage of horizontal waves; this happens because light that strikes a shiny surface at the special angle of 57° (called Brewster's angle) is filtered or polarized by the surface so that only waves that vibrate parallel to the surface bounce off and reach your eye. In the case of a car hood, horizontal waves reach you since the hood is horizontal.

Polaroid sunglasses contain vertical ribbing as fine as the height of a single light wave—about a millionth of an inch—which polarizes light vertically, stopping all but the vertically vibrating waves. This wipes out glare from horizontal surfaces because the horizontal light waves coming off at Brewster's angle can't fit through the Polaroid ribbing, which runs crosswise to them. Most things that are important for us to see move on horizontal surfaces, such as roads, floors, and bodies of water, so a vertically polarizing lens takes out the worst glare. We see clearly through the lens because there are enough vertical waves that do get through to light the scene for us.

10. Spaceship Earth

1. They are all true!

(a) A rainstorm occurs in an area of low barometric pressure. When there is less air pressure on your body, the gases in your joints expand and cause pains;

(b) A storm is often preceded by humid air. Frogs have to keep their skins wet to be comfortable, and moist air allows them to stay out of the water and croak longer;

(c) A low-pressure rain system moving into an area will often stir up a south wind that flips leaves over;

(d) Ice crystals form in high-altitude cirrhus clouds that precede a rain front. These crystals refract light from the moon and make a ring around it;

(e) Birds' and bats' ears are very sensitive to air pressure changes, and the lower pressure of a storm front would cause them pain if they flew higher where the pressure is even lower;

(f) Cold-blooded crickets chirp more the hotter it gets. Count the number of chirps a cricket makes in 15 seconds and add 37—this will give you the temperature in degrees Fahrenheit;

(g) Rising humidity causes the ropes to absorb more moisture from the air and shrink;

(h) Fish come up for insects which are flying closer to the water because of lowered atmospheric pressure;

(i) A rising wind, often marking the coming of a storm, causes a high whining sound when it blows across telephone wires.

2. Water vapor is the culprit! The earth, being a relatively cold body, radiates most of its energy in the longer infrared wavelengths

(4 − 70 · 10⁻³ mm). It so happens that water vapor is an excellent absorber of infrared radiation. As a result, even on perfectly clear nights, a humid atmosphere greatly restricts the radiation of heat to space, and the lowest nightly temperatures occur with very dry air.

3. The inshore part of each wave is moving in shallower water. There the friction of the bottom causes it to slow down. Thus the inshore part moves slower than the part in deep water. The result is that the wave front tends to become parallel to the shoreline (see diagram). We can also see that this process has the effect of concentrating wave energy against headlands. It is a modern expression of the old sailor's saying, "The points draw the waves."

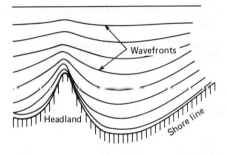

4. This is an example of a temperature inversion where the temperature increases rather than decreases with height. On clear nights with little wind inversions are common because the ground loses heat by radiation more effectively than the air above, resulting in rapid cooling of the ground at night. Lack of wind helps for it prevents the mixing between the cooler and the warmer layers of air.

5. If we take a compass needle and pivot it so that its ends can move up or down, we'll see that the north end will dip about 60° to 70° from the horizontal. One look at the globe will convince us that the north end is simply pointing along the shortest route through the earth to the magnetic pole in northeastern Canada. Similarly, the magnetic domains in stationary iron objects turn around until they line up with their north-seeking ends pointing 60° to 70° downward, and south-seeking ends pointing directly opposite. The combined effect of millions of such magnetic domains all pointing in the same direction pro-duces a magnetic north pole at the bottom and a south pole at the top.

6. Gas, in this case the water vapor, releases energy in transforming to the liquid and/or solid state. Conversely, a solid must absorb energy to melt and a liquid must absorb energy to vaporize.

7. Airliners have pressurized cabins. The air at a height of 30,000 feet is compressed to sea level pressures. This process would raise the temperature of the air to 130°F if air conditioners were not used to extract heat from the air.

8. The effect may be due to the Coriolis force which is about 50% stronger at the poles than in middle latitudes. A walking person makes corrections for the Coriolis force easily and quite unconsciously. On frictionless ice that prevents his making any small lateral corrections (but somehow still permits him to walk!) a person walking at 4 mph would drift from his intended straight path by about 250 feet at the end of one mile. It is said that even the penguins in the Antarctic waddle in arcs to the left, although the author cannot guarantee the scientific ac-curacy of this statement.

9. False! If winds rushed directly toward areas of lower pressure, no strong "highs" or "lows" could develop, and our weather would be much less changeable than it is. However, due to the Coriolis force caused by the rotation of the earth, wind from any direction veers to the right in the Northern Hemisphere. As a result, the whole air mass initially flowing directly toward a low-pressure area begins to rotate counterclockwise (see diagram). This in turn tends to slow the filling of the low-pressure region since now the pressure difference supplies a centrifugal force which tends to keep the winds moving in circular paths. In the Southern Hemisphere the Coriolis force causes winds to veer to the left, and so the direction of circulation is clockwise.

Near the equator the Coriolis force is zero or very small. In that re-gion any atmospheric pressure differences produced by heating of the air at the ground are quickly smoothed out, and the region has well earned the name of "the doldrums." Hurricanes and typhoons rarely form closer to the equator than 5° latitude.

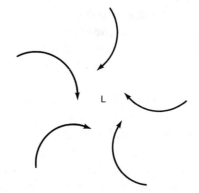

10. 1. The direction the observer was facing can be deduced by examining the situation at two other latitudes. At the North Pole, the elevation of the sun would be nearly constant during the day. At typical latitudes in the lower 48 states, the sun reaches its highest elevation when it is due south and its lowest upon rising in the east and setting in the west. As we move farther north we expect the rising and setting locations to move farther north until they meet due north of us. Therefore the observer was facing north.

2. The sun reaches its highest elevation when it is due south. This time is known as local noon. The lowest elevation is then obtained at local midnight.

11. There are three different ways to look at the origin of meanders.

The first is the mechanical model. Assume that a small bend of a river comes into being due to some minor irregularity of terrain. The centrifugal force that arises as the water goes around the bend tends to fling the water outward toward the concave bank. Because the water at the top surface of the river is slowed less by the friction of the river bed, it moves across the stream toward the concave bank and is replaced from below by water that moves across the bottom of the stream in the opposite direction (see diagram [a]). The concave bank is scoured by the downward current and eventually eroded, thus increasing the sharpness of the bend. This throws the river into a path that traverses the hill rather than coursing straight down. Eventually, however, gravity pulls the river around into a downhill path, creating an opposite bend. Thus the process continues.

Looking at meanders from a different point of view, they appear to be the form in which a river does the least amount of work in turning. Clearly, work is required to change the direction of a flowing liquid. The work is minimized if the shape of a river has the smallest total variation of the changes of direction. This property can be demonstrated by bending a thin strip of spring steel into various configurations by holding the strip firmly at two points and allowing the length between the fixed points to assume an unconstrained shape (see diagram [b]). The strip will assume a shape in which the direction changes as little as possible. This minimizes the total work of bending since the work done in each element of length is proportional to the square of its angular deflection. The bends are not circular arcs, parabolic arcs, or sine curves; they are special functions known as elliptic integrals.

The third model for meanders comes from analyzing the course of a river in terms of randomness and probability. One can prove that any line of fixed length that stretches between two fixed points is likely to follow a meander. The proof consists of generating random walks or paths in which a moving point can strike off in a direction determined by some random process (for example, the throw of a die or the sequence of a table of random numbers), as it journeys between two fixed points in a specified number of steps. The most probable path for such a moving point is a serpentine pattern with proportions similar to those found for rivers. It is a paradox of nature that such random processes can produce regular forms, and that regular processes often produce random forms.

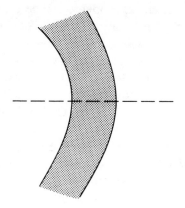

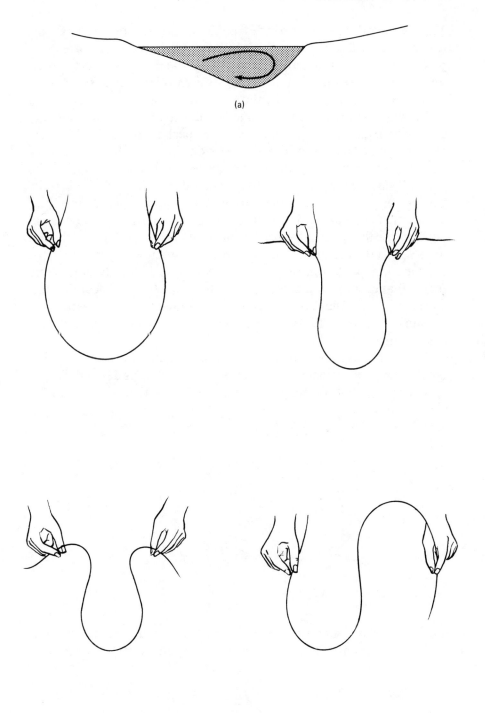

(a)

12. The earth's rotation provides us with a regular sequence of events that can be used to define a unit of time. The simplest sequence is the passage of a given star across the meridian, the imaginary north-south line drawn across the sky and passing directly overhead. A star rises in the east, crosses the meridian, and sets in the west. The time measured in this way is known as sidereal time. A sidereal day corresponds to exactly one rotation of the earth with respect to the "fixed" stars. However, as a measure of time for everyday life the sidereal day is not too practical because we are far more affected by the sun than by the "fixed" stars. Thus we are led to the solar day which is defined as the period between two successive transits of the sun across the local meridian. The solar day is about four minutes longer than the sidereal day because the earth has to rotate a little bit more than 360° for the sun to return to the meridian (see diagram). The picture shows the earth's orbit as circular, but in fact it is slightly elliptical with the sun in one focus. Planets in elliptical orbits go faster when they are near the sun, and slower when they are farther away. This means that the earth near the sun has to rotate farther to get the sun back to the meridian than it has to when it is farther away.

The perihelion, point of closest approach to the sun, occurs during northern winter. Thus the days in northern winter, as defined by successive meridional transits of the sun, are longer than they are in the summer.

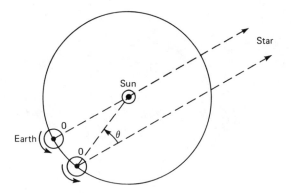

13. The land masses today happen to be distributed in such a way that a continent lies astride the South Pole, and the North Pole, though oceanic, is almost landlocked. Both of these features make it impossible for enough warm water from the tropics to reach the poles and inhibit ice formation. The present-day polar ice caps are therefore a direct result of continental distribution. The plate tectonics theory suggests that in earlier geological ages the continents were distributed differently, thus inhibiting the formation of polar ice caps.

14. The astronomical reason is the earth's elliptical orbit. At perihelion, the point of the orbit nearest the sun, the earth is 1.407 $\times$ 10⁸ km from the sun. At aphelion, the point farthest from the sun, the earth-sun distance is 1.521 $\times$ 10⁸km. There is relatively little difference but it is not negligible. Happily for the northern hemisphere, perihelion occurs during winter, and this helps to moderate the seasonal effect produced by the tilt of the earth's axis to the orbital plane. The reverse is true for the southern hemisphere, which would suggest that the latter should have colder winters and hotter summers. However, the greater area of ocean in the south serves as a moderating influence. The high heat capacity of water means that in summer the ocean is slow to warm and in winter slow to cool. This makes summers in the southern hemisphere somewhat less hot and winters less cold than they would be otherwise.

15. It is true that near the ground the highs are generally cold and the lows warm. However, for higher altitudes we must consider the variation of pressure and density with height. Due to gravity most of the atmosphere is concentrated near the ground. The reason why all of the atmosphere does not collapse completely is that the downward gravitational pull on each parcel of air is balanced by the upward push due to the higher pressure from below. This happens if the pressure and density of the atmosphere decrease exponentially upwards. The exact formula is $P = P_o \, exp \, (-hg/RT)$, where h is the height and P_o is the pressure at ground level. We see that pressure falls off with height more slowly in warm air than in cold air (see diagram). As a result, at any given height pressure is higher in the warm zone than in the cold zone. This horizontal pressure difference grows with height and generates the thermal wind. For example, the thermal wind associated with the polar-subtropical temperature difference is always on average westerly and manifests itself as the circumpolar jet stream which snakes around the pole in a wavy manner. In frontal systems the thermal wind consists of a high-altitude jet running along the line of the front. Sometimes these upper winds are made visible by cirrus clouds and their direction can be seen by the cloud motion.

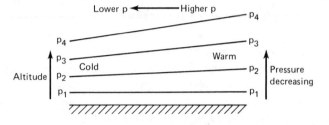

16. We might think that a mountain range could be represented by a long half-cylinder of density d_m lying on a flat plane (see diagram, [a]). This model, however, predicts angles of deflection that are much larger than what is actually observed. Suppose instead that the mountain range can be represented by a long cylinder of density d_m *floating* in a fluid of density $2d_m$ (see diagram, [b]). Then one can show that the plumb-bob deflection due to the mountain range is zero in this model. This makes good physical sense: The mass contained in the top and bottom halves of the cylinder is exactly the same as the mass of the earth that would be found in the bottom half of the cylinder if the mountain range weren't there. The success of this model has convinced geologists that mountains, and also continents, float on the underlying mantle rock.

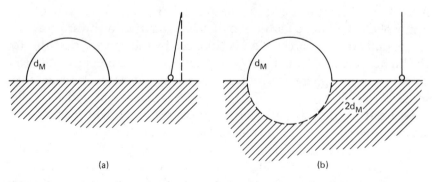

(a) (b)

17. The train going west. Its speed subtracts from the west-to-east rotation of the earth, whereas the speed of the train going east adds to it. This results in a smaller centrifugal force subtracting from the weight of the train traveling west, thus making it heavier.

18. Astronomical nightfall is defined as the time when the sun sinks 18° below the horizon. In the tropics the sun follows a much steeper path through the sky than in the subpolar regions, and thus sinks faster below the horizon. For example, at equinox (March 21 and September 23) night begins 1 hour 12 minutes after sunset on the equator (e.g., Quito, Ecuador), 1 hour 44 minutes after sunset at a latitude of 45° (e.g., Portland, Oregon), and 2 hours 33 minutes after sunset at a latitude of 60° (e.g., Anchorage, Alaska). We can also see that nightfall is more rapid in summer regardless of the latitude.

11. Space Exploration

1. This location was selected because of the land-free ocean extending 5000 miles to the coast of South Africa. This is important because it makes it possible for the first two stages of three-stage rockets launched over the Atlantic to fall into the water with little chance that they will fall on populated areas.

Why choose an East coast launching site instead of a West coast one? Earth's rotation provides the answer. A rocket on the ground at Cape Canaveral is being carried eastward at 910 mph. This speed is calculated by dividing the distance around the earth at the latitude of Cape Canaveral (28.5°N)—21,800 miles—by 24 hours. A satellite in circular orbit must move at 17,300 mph. If it is already going 910 mph on the ground, the additional velocity required is only about 16,400 mph. Conversely, if the rocket were to be launched toward the west, the propulsion system would have to give the vehicle a velocity of about 18,200 mph in order for its westward speed to be 17,300 mph.

2. No. The gravitational force on the satellite is always directed toward the center of the earth. Hence the plane of its orbit must also pass through the center of the earth. For a satellite to hover over New York, however, it would have to always move east at the same velocity as New York is rotating with the earth. But such an orbit would not contain the center of the earth, so it could not be realized.

Even though it is not possible for a satellite to hover directly over New York, it is perfectly feasible to launch a satellite so that it is seen at the same point in the sky from New York. Place it in a west-to-east orbit above the equator with a mean height of 22,300 miles, i.e., about 5½ times the earth's radius. The satellite will have a period of 24 hours.

This fact is used in launching communications satellites. Three equally spaced satellites placed in geosynchronous orbit can cover the earth.

3. Spin a coin on the floor of your room. The coin will refuse to spin because, by conservation of angular momentum, a spinning object tries to maintain its position in space, while the space station is continually changing its position in space.

4. Yes. The total energy of a rocket of mass m and speed v on the surface of the earth of radius R is $1/2mv^2 - GMm/R$. The first term is the kinetic energy of the rocket, and the second term is its negative potential energy in the gravitational well of the earth. To escape from the earth, the rocket must have enough kinetic energy to overcome its gravitational pull, i.e., $1/2\ mv^2 = GMm/R$. This expression is independent of the direction of v, so it doesn't matter which way the rocket is pointing.

5. Harder, because sidewalks would be more slippery. The friction between the shoes and the underlying surface is proportional to the person's weight, which on the moon is only one sixth of the weight on the earth.

6. The launching rocket is generally larger than the satellite. As a result it encounters more air resistance, and slowly loses altitude. In so doing, the rocket converts its potential energy into work against air resistance and an increased kinetic energy, i.e., a greater speed. Thus the increased speed follows from the principle of the conservation of energy.

7. Before a space vehicle is launched, it already is traveling with the earth around the sun at about 30 km/sec. To leave the earth, the vehicle must reach a velocity at burnout (the moment when its rocket motor stops firing) of at least 11 km/sec with respect to the earth, assuming that its altitude at burnout is at least 200 miles. However, by the time the vehicle has coasted "uphill" out of the earth's gravitational field, it will have slowed down to about 3 km/sec. If all or part of this residual velocity is added to the 30 km/sec solar velocity by launching the vehicle in the direction of the earth's motion, the vehicle will move ahead of the earth and out on an elliptical course toward Mars. If the 3 km/sec residual velocity is subtracted, by launching the space vehicle against the direction of the earth's motion, it will fall behind the earth on an elliptical course toward Venus.

8. In the free-fall condition that prevails while a spacecraft is in orbit, weight cannot be measured directly, but mass can. This is made possible by a very basic principle: The frequency at which an object supported by a spring vibrates depends on the mass of the object. Skylab was equipped with a chair mounted on springs. A man would seat himself in the chair, strap himself in, and set the chair vibrating. The chair had built into it equipment with which he could accurately time the period of the vibrations. From this he could calculate his mass, which is directly proportional to his weight as measured on earth.

9. Yes. This paradoxical fact can be understood by realizing that the exhaust gases always come out at the same velocity relative to the rocket, while the latter is constantly accelerating. Obviously at some point the rocket's forward velocity will exceed the gases' backward velocity, and relative to the ground they will start moving forward. Mathematically speaking, one can derive an equation for the velocity v of a rocket at any given time t as a function of the initial mass m_o of the rocket, the mass m of the rocket at time t, and the velocity u of the exhaust gases with respect to the rocket. The equation is simply $v = u \log_e (m_o/m)$. It is easy to see from this that as soon as the rocket has burned fuel to the point where $m_o/m > e$, v becomes greater than u, and with respect to the ground the exhaust gases travel in the same direction as the rocket.

10. The rocket experts are right. The parabolic path studied in elementary mechanics is derived under the assumption that the gravitational force acting on the projectile acts in the same downward direction throughout the journey of the projectile. This amounts to assuming that the earth is flat. In reality the force is always directed toward the center of the earth, and that direction varies from point to point. Therefore, a stone does move in an elliptic path with the center of earth at one focus. However, for small speeds the eccentricity of the ellipse is close to 1 (i.e., the ellipse is highly flattened). This implies that small segments of the ellipse from the point of throw to the point of touch-down may be considered parabolic (see diagram).

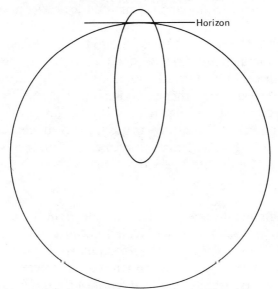

11. Weightlessness can be achieved when a pilot flies a carefully controlled parabolic course (see diagram). The centrifugal force (dashed arrow) is then exactly balanced by the gravitational attraction of the earth (solid arrow). A projectile thrown in the gravitational field of the earth also describes a parabolic path, and is weightless. If this seems hard to believe, make a hole in the bottom of a can full of water, and throw it at an angle to the ground. No water will be flowing out of the can while the latter is in flight!

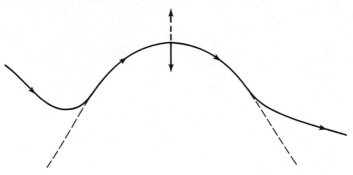

12. During the lift-off and reentry phases of space flight astronauts experience accelerations of 10–20 g for periods of many minutes. (1 g is an acceleration in which the speed increases or decreases by about 10 m/sec during each second, and is equal to the acceleration of a freely falling body. In British units, 1 g is equivalent to accelerating from 0 to 60 mph in about 3 seconds.) It was found that the effect of accelerations on the human body varies depending on whether the astronaut is lying in the direction of the acceleration so that his blood is forced from the head to the feet, or whether he is lying in a prone position so that his head and heart are at the same relative level as far as the accelerating forces are concerned. In a sitting-up position loss of consciousness occurs already at 4–8 g, whereas in a prone position the astronaut can tolerate up to 17 g for short periods of time without losing consciousness.

13. All the landing sites were on the near side of the moon, not because there was no interest in the far side, but because of the need to communicate with the earth by radio. Astronauts would not have been able to send and receive radio messages on the far side since electromagnetic waves could not travel through the interior of the moon in order to get to the earth.

12. The Universe

1. It is far easier to escape from the sun completely than it is to get to the sun. The earth travels around the sun at a speed of about 66,000 mph. A spacecraft from the earth would only have to go 28,000 mph faster to escape from the solar system. But to drop inward and reach the sun itself, the spacecraft would have to slow down drastically *with respect to the sun* by accelerating the nearly 66,000 mph in the direction opposite to the motion of the earth. The entire earth would have to do likewise before it could begin to fall into its parental star.

2. The lunar surface is full of craters, mountain-walled plains, and other irregularities. These surface features cast long shadows when illuminated obliquely by the sun, as during the first or last quarter. The shadows make the surface appear darker than at full moon when the sun shines directly from above over most of the lunar surface. Note that due to the eccentricity of the moon's orbit around the earth one full moon is not equal to another! The distance to the moon varies from as little as 356,400 km to as much as 466,700 km, and accordingly light of full moon can vary by as much as 30%.

3. No. The moon's rotation has become synchronized with its revolution about the earth. As a result, the same hemisphere of the moon is always turned toward the earth. Thus, to an observer at a given site on the moon the earth will always appear at the same point in the sky. For example, from near the center of the visible lunar hemisphere the earth will be visible directly overhead, although it will oscillate a little about this position due to the tiny swaying of the moon called libration. Of course, the earth will be seen to go through the phases the way the moon does from the earth.

4. No. A solid disk obviously doesn't move like this. Looking at a spinning phonograph record, we can see that the outer edge moves much faster than its interior points. But just as Venus moves faster than the earth under the influence of the sun, so objects closer to Saturn move faster than those farther out. The conclusion is that the rings are made of swarms of small particles, ranging in size from dust to huge boulders, each in orbit around Saturn.

5. As Venus revolves within the earth's orbit, its sunlit hemisphere is presented to the earth in varying amounts. It shows its full phase at the time of superior conjunction (see diagram), the quarter phase on the average near the elongations, and the new phase at inferior conjunction. Paradoxically, Venus is at its brightest not when it is nearest the earth (its new phase), but in its crescent phase (about 5 weeks before and after the new phase). On the other hand, the earth, being farther away from the sun than Venus, presents all of its illuminated hemisphere toward Venus when the two planets are closest.

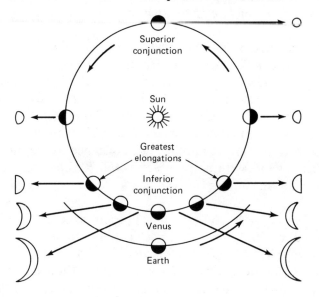

6. Mercury and Venus follow orbits between the earth and the sun. As a result, to an observer watching the sky they are never far from the sun, their maximum angle from the sun being 28° for Mercury and 48° for Venus. Our observer scanning the sky from a part of the earth where night has fallen is looking away from the sun. Mercury and Venus lie behind him toward the sun, and are therefore absent from the night sky.

Mercury is as bright as the brightest of the stars, but can only be seen for a week or two three times a year in the evening and three times before sunrise. Venus sometimes remains in the sky up to four hours after sunset, but is best visible as a "morning star" or an "evening star." It is interesting that Venus, like the moon, can on occasion be seen in full daylight, and warships have even been known to fire at it, mistaking it for an enemy balloon.

7. The morning side of the earth is struck by both the meteors it encounters, and also those which it overtakes, while the evening side is hit only by those meteors that gain on the earth (see diagram).

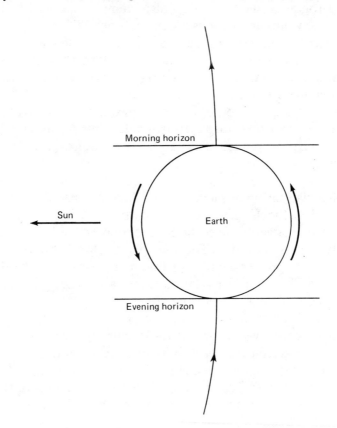

8. The gravitational field of the massive planets is high enough to attract and hold considerable atmosphere. The gases of such an atmosphere are low in density compared with the rocky main body of the planet, and their presence greatly reduces the average density of the planet as a whole.

9. There are quite a few! One is the inner satellite of Mars, called Phobos, which revolves around Mars in 7 hours 39 minutes. This is less than one third of the rotation period of the parent planet. As a result the easterly orbital motion of Phobos in the Martian sky far outweighs its apparent westerly motion caused by the rotation of Mars, thus making it rise in the west and set in the east.

Another object is the sun as seen from Venus or Uranus. Viewed from the North Star, all planets revolve around the sun counterclockwise and rotate around their axes also counterclockwise, i.e., from west to east. Venus and Uranus are the only exceptions: For example, Venus turns from east to west on its axis, and extremely slowly at that. Its day is equal to 243 earth days. The retrograde rotation of Venus has, of course, the effect that the sun there rises very slowly in the west and sets just as slowly in the east.

In addition, Jupiter's outer four satellites, Saturn's moon Phoebe, and Neptune's moon Triton have retrograde orbits around their parent planets which perhaps indicates that they have been captured from the neighboring planets.

10. If the relationship displayed by the data obtained from the other planets held for the earth, it would be rotating in 15.5 hours rather than in 24. However, over the ages the earth's rotation has been slowed by the tidal effects of the moon. The other planets, Mars, Jupiter, Saturn, Uranus, and Neptune do not have any satellite as large in relation to themselves as the moon is in relation to the earth. They therefore have not suffered a comparable slowing effect.

The moon itself suffers a slowing effect even greater than that sustained by the earth. While the earth is affected by the moon's gravity, the moon is affected by the earth's 81-fold greater gravitational field. The moon's rotation has been slowed to a complete standstill with respect to the earth, so that one of its sides always faces us. Its rotation with respect to the sun, however, has not stopped. Its solar day is about

29 1/2 earth days, which is equal to its period of revolution about the earth.

The period of rotation of Mercury has been drastically slowed by the sun's tidal effects and is now equal to 88 days, a period equal to its time of revolution about the sun. Venus has also been slowed down by the sun, and now takes 243 days to rotate on its axis, which is close to its period of revolution about the sun (225 days).

11. No, because inside the crescent illuminated by the sun is the dark region of the moon, which hides from view any stars that might be in that part of the sky.

12. Yes. The moon (and even the sun) may appear blue after large forest fires or volcanic eruptions. The apparent color of the moon or the sun is to a large extent determined by the scattering of the light coming from them by the atmospheric molecules. Light with shorter wavelengths (the blue end of the spectrum) is deviated more from its original direction than light with longer wavelengths (the red end). Thus the setting sun looks red, since it is depleted in the blue color, and the sky looks blue because what we see as the sky is mostly the scattered blue light. However, particles produced by the fires or volcanoes are often the size of the visible light wavelengths (around 10^{-4} cm), and paradoxically scatter the red end of the spectrum more than the blue. As a result an observer sees the moon depleted in red, i.e., left with the blue end of the spectrum.

13. A mountain cannot rise higher than a certain critical height which on the earth is about 90,000 feet. Any greater height would increase the weight of the mountain to the point where its base would start turning into a liquid under such enormous pressures, thus causing the mountain to sink below the critical height. On Mars the gravitation is less than on the earth, the mountains therefore are lighter, and thus can reach greater heights.

14. The apparent orientation of the moon's surface varies widely depending on the observer's latitude, and for a given latitude on the position of the moon in the sky. Thus the lunar mountains (light areas) and maria (dark areas) can appear in vertical, horizontal, reversed, and all other intermediate positions depending on where you are on the earth. If you take two observers along the same meridian, say one in Boston and the other in Santiago, Chile, the observer in Chile will see the moon exactly upside down compared to his friend in Boston only when the moon is due south. At other times the relative orientation is more complicated.

15. Not very much because weight is defined as the force of the earth's pull. The moon's distance from the earth, R, is so large that the expression for the force of the earth's pull

$$W_{moon} = GM_{earth}M_{moon}/R^2$$

has an extremely large denominator, thus making W_{moon} rather small.

16. Yes! The distances between the individual stars within a constellation appear to increase when the constellation is close to the horizon. This effect is particularly striking for the constellation of Orion in winter, and Cygnus in summer.

Index

Index